The Spread Mind

The Spread Mind

Why Consciousness
and the World
Are One

Riccardo Manzotti

O/R

OR BOOKS

NEW YORK • LONDON

To my father Bruno and my mother Jolanda, who have always believed in the power of ideas.

© 2017 Riccardo Manzotti

All rights information: rights@orbooks.com

Visit our website at www.orbooks.com

First printing 2017.

All rights reserved. No part of this book may be reproduced or transmitted in any form or by any means, electronic or mechanical, including photocopy, recording, or any information storage retrieval system, without permission in writing from the publisher, except brief passages for review purposes.

Library of Congress Cataloging-in-Publication Data: A catalog record for this book is available from the Library of Congress.

British Library Cataloging in Publication Data: A catalog record for this book is available from the British Library.

Text Design by Pauline Neuwirth

Published for the book trade by OR Books in partnership with Counterpoint Press.

Distributed to the trade by Publishers Group West.

hardcover ISBN 978-1-94486-949-6

ebook ISBN 978-1-94486-956-4

CONTENTS

INTRODUCTION

IN *THE SPREAD MIND*, I present a view of nature that challenges the traditional separation between appearance and reality, between experience and objects, and between mind and nature. I discuss and revise key notions—such as existence, experience, appearance, consciousness, representation, relation, causation, identity, and the now. This endeavor moves from a simple idea, namely that the thing that is my conscious experience is not an inner ghost but the very object I am conscious of. Perhaps surprisingly, my experience of an apple is the apple itself. The subject is the object. How is this possible? The proposed solution is that the separation between a physical object and a mental world was the offshoot of an oversimplified notion of physical objects. If we understand that the external objects that populate our life do not exist autonomously but only relative to our bodies, we will no longer need to place our experience in an inner mental domain. Our experience will be one and the same with the physical objects that surround our bodies. Objects are not absolute, though. They are relative and, as a consequence, they are spread in space and time as our experience is—the two being the same.

Such a hypothesis does not rely on armchair efforts, but derives from the wealth of data that psychology and neuroscience have accumulated in the last decades. To make a convincing case for the spread mind, I will revisit the empirical evidence about perception, illusions, hallucinations, and dreams. I will show that all cases of mental experience can be explained as cases of perception.

Philosophers and scientists have often assumed that one's experience has—at least in principle—only a contingent link with the real world: on the one side physical reality and on the other side our subjective qualitative untrustworthy experience. Yet, perhaps surprisingly, a different picture will be drawn. I will show that experience and reality are one and the same. If that is the case, experience—and thus consciousness—will fit seamlessly with nature. Experience will not be an addition to the physical world, but will be a subset of it—tantamount to how living organisms are a subset of the physical world. This insight is here elaborated upon and expanded into an original and comprehensive theory of both consciousness and physical objects—the theory of the spread mind.

The key intuition—namely that *one's experience of an object is the object one experiences*—suggests a new perspective about consciousness—namely that *all experience is perception of actual physical objects*. The book starts from the premise that nature is all there is. Hence, if nature is made of objects, experience must be made of objects as well. Therefore, articulating the premise of the spread mind has a twofold goal. It explains what experience is made of and it revises the notion of the object in the physical world. The object is no longer a passive entity but it is instead an active cause.

Up to now, the reduction of experience to neurons and their whereabouts, the usual candidates for the physical underpinnings of consciousness, has not been satisfactory for explaining the conscious mind, since experience and the brain do not resemble each other in the least. In contrast, I suggest a radically different account: conscious experience is not an exotic phenomenon concocted by the brain, but it is the subset of the physical world one experiences at any moment. Such a subset is not made of neural firings inside brains but of physical objects outside one's body. Consciousness is physical, and it is outside one's body. Our mind is physical and yet, ironically perhaps, is neither our body nor our brain (or any property of them).

The view proposed in this book allows us to shed a new light on the vague notion of personhood, too. Persons are not their bodies. A person is a set of objects and events spread in space and time. *A person is a world that exists causally relative to his/her body. The body is the proxy that allows the objects one experiences to produce effects here and now.* In this regard, a person is a physical part of the world, like a pebble or a thunderstorm, only the world is not identical with one's body.

If experience and physical objects are the same, the proposed solution can address another thorny notion—namely, the relation between mind and world. In nature, one cannot observe any relation. No picture of relations has ever been shot. No relation has ever been measured. Relations—such as intentionality, semantics, and representations—have never been observed experimentally. They have been postulated to fit conceptual gaps between cherished theories and everyday life, but they are not found in the world as it happens, say, to stars, trees, and electrons. In this regard, the spread mind suggests a radical move. There are no relations of any kind in the world, there are only objects and everything is just identical with itself. If one's experience of the world is the world one experiences, no dubious relation is needed. In the physical world, *only one kind of relation is needed, and indeed available—namely, identity.* Once we relocate ourselves in the world, we no longer need any bridging relation. We are already there, so to speak.

The presented view does not require customized ontological licenses, rather it rolls back entrenched conceptual distinctions and thereby flattens our view of nature to a uniform landscape of objects causally related. Once the alleged gap between appearance and reality is set aside, nature and experience will not need separate levels of description. The traditional separation between what things are and how they appear can be set aside. The goal is to show what and where experience is within the physical world—what and where consciousness is in nature. The contrasting notions of appearance and reality might be akin to those of the directions left and right—i.e., useful, but not essential, ways to describe reality.

The good news is that, if the theory of spread mind has any merit, it will pave the way to a new interpretation of the existing empirical findings. A lot of puzzling data from neuroscience will become amenable to a new interpretation. Neural firings and neural structures will no longer have to carry the burden of creating an inner mental world.

No special property will have to emerge out of neural activity. The brain will no longer be a unique case in which something unexpected and unexplained occurs. Neuroscience will be able to integrate its findings in a broader context that, without looking for mysterious mental properties, will locate the physical underpinnings of consciousness in the objects that brains bring into existence.

If I were hard-pressed to explain why consciousness has become an almost hopeless mystery in both science and philosophy, I would not hesitate to say the culprit has been—and still is—the glorification of misinterpreted empirical evidence in terms of dogmatic prejudices lacking empirical confirmation. Cases of misperceptions, such as dreams or hallucinations, have been transmogrified into metaphysical truths backed up by modal logic and its cognates. Because of the alleged difference between experience and reality, philosophers and scientists have drawn the conclusion that there is an unbridgeable gap that looms between what the world is and how it appears. Yet, such a conclusion is questionable because it requires independent knowledge about the thing that is our experience and the world. In other words, how can we state that experience does not look like physical things— whatever physical things are—if we do not know what experience is? Why should we rule out the much simpler hypothesis that experience is one and the same with external objects?

Moreover, a persistent confusion between experiencing something and knowledge about what one experiences has muddled matters further. In this regard, this book will focus on the issue of experience only—i.e., consciousness. Is experience ever independent of the physical world as many philosophers have often suggested? As a matter of fact, the Cartesian doubt—i.e., the possibility that what we experience has no connection with physical reality—was not based on metaphysical intuitions but on the empirical evidence available in the seventeenth century. Consequently, by revising empirical data, the separation between consciousness and the physical world can be reconsidered. A goal of this book is to show that, although we can have misbeliefs, we cannot *mis*-experience the world. Whatever we experience is of a piece with nature and, as such, it is like a rock: it cannot be a mistake. It can only be. A rock is a rock. An apple is an apple. An apple is never a mistake. Likewise, if experience is one and the same with reality, experience is never a mistake. Nature is never a mistake. I can see heated sand and misjudge it to be a pool of water, but I cannot mis-experience something. If I see red, something must be red. If I see a red apple, a red apple must exist.

The idea of the spread mind bridges the gap between mind and world by undermining the contrast between appearance and reality. The contrast arises in the first place by assuming that experience is an outcome of neural activity. After all, is there any difference between a red apple and our experience of it? In the same spirit, the view recasts dreams and hallucinations in terms of the perception of physical objects like those we meet in everyday life. The difference between everyday life and dreams is more a matter of degrees than the result of an unfathomable chasm in the fabric of reality.

The main premise of this book is that experience, since it is real, is a part of nature. I am real. My conscious experience is real. Thus consciousness must fit in nature as everything else. Nature is not a club that does not want me as a member. Nature is the totality of the physical world. Whatever exists partakes of it. Perception, dreams, and hallucinations are as much a part of nature as pebbles, rivers, storms, and stars. The very idea—ever popular—that experience is an invisible private citizen of an inner phenomenal/mental world is irredeemably suspicious. Why should all natural entities but experience be observable? Why should experience be special? If the received view suggests that experience is different from nature—being private, qualitative, and relational—so much the worse for such a view. Experience is a part of nature and, as such, must be like everything else. Experience must fit inside nature or, symmetrically, nature must encompass experience.

If experience is real, there will be a physical phenomenon that is identical to it, in the same way that if electricity is real, electrons will exist somewhere. Such an obvious idea has been blocked by two misleading notions: on the one hand, the idealized object of physics, and on the other hand, the philosophical notion of experience. In contrast, the key notion of the spread mind is that *the red apple I experience is my experience of the red apple*—namely, a space-time *spread* object. I claim that such an object encompasses all cases of misperception. I will show that misperception, dreams, illusions, and hallucinations are not different from perception and, in turn, not different from everyday objects. By actual, standard, or normal perception I will refer to cases in which one perceives something that is there.

Various arguments, both empirical and conceptual, will address squarely the distinction between appearance and reality, between the mental and the physical, and between experience and objects. The aim of such arguments is to show that, in contrast with venerable traditions, dreams and hallucinations are not mental stuff but are instead physical spatiotemporal causally active relative objects—and yet objects nonetheless.

The pillars that support the theory of spread mind are a causal notion of existence, a solution to the thorny issue of representation in terms of identity, and a causal notion of the now. In sum:

- All experience is perception
- All perception is identity with a physical object
- All objects are causally singled out
- All objects are relative and intrinsically temporal
- All objects are spread objects

Something of what follows might appear simplistic to the astute and sophisticated scholar who is accustomed to subtle conceptual distinctions. However, a certain degree of coarseness is both welcome and unavoidable if a radical departure is attempted. The goal is a conceptual reboot capable of spawning empirical predictions amenable to falsification. The endeavor is both conceptual and empirical and will be matched against the available empirical evidence from neuroscience, cognitive science, and psychology.

Since Descartes, appearance and reality have been assumed to be distinct. In the wake of Kant's work, the world has become unknowable. In the 1950s, the findings of Canadian neurosurgeon Wilder Penfield convinced many that we perceive reliable hallucinations concocted by our brains. A successor to Penfield, English physiologist David Marr, held that only an indirect model of the external world is available to one's mind. Many contemporary neuroscientists believe we see a virtual world generated by neural machinery. In both science and philosophy, most models are consistent with the quite depressing belief that one's mind is always one-step behind the world. Or, at least, this is the bleak view that most scientists and philosophers are eager to adopt. While the popular notion that the mind is different from the external world shelters us from the matter, it also secludes us from an intimate contact with the nature of the world. The mind never enters or peers into the *sancta sanctorum* of the physical world. Only appearances and shadows are offered to our own limited epistemic powers. While psychoanalytic explanations might support such a recurring abstinence from actual intercourse with the very flesh of the world, a different explanatory route ought to be taken seriously. In contrast to such views, the stuff our experience is made of might be the very physical world we experience. Consciousness might be spread in the world beyond the boundaries of one's body. Let us check whether the time has come to bring down the curtain on this *Tristan and Isolde* drama of subject and object, mind and world, philosophy and science.

1.

The Spread Mind

I am what is around me.

—WALLACE STEVENS, 1971

NE'S CONSCIOUS EXPERIENCE of an object is the object one experiences. Until now, this unexpected game-changing solution to the mind-body problem has not received the serious consideration that it deserves. When, say, Emily looks at a red apple on a table, the physical underpinning of Emily's conscious visual experience is nothing but the red apple sitting on Emily's table, outside of her brain and outside of her body. Such an object is as physical as Emily's brain. Neither empirical evidence nor known laws of nature forbid Emily's experience from being one and the same with the red apple on the table. Both apples and neurons are physical. If the conscious experience of a juicy red apple can coalesce out of neural firings, why could it not coalesce out of the juicy red apple itself? In fact, apples are much closer to one's experience of apples than neurons are. A caveat: in these pages, I will take the words *experience, mind,* and *consciousness* to be synonyms.

The view that *one's experience of an object is the object one experiences* is called the spread mind,[1] and this book explores the many advantages of this way of thinking over the standard view. The standard view holds, give or take, that one's experience of the world is different from the world one experiences. Experience—i.e., consciousness—is assumed to be a property of neural activity if not neural activity itself. Locating experience inside the body leads to countless unsolvable mysteries. How can experience reach and grasp the external world? How can phenomenal properties match against physical properties? What is the relation between consciousness and the physical world?

The idea of the spread mind resolves in a simple way the issue of phenomenal versus physical properties—it says they are the same. However, acceptance of this theory does not come cheaply. The scientist will be forced to consider the brain as a part of the world, rather than as the organ that secretes consciousness, much as the pancreas secretes insulin. The philosopher will have to give up a carefully balanced house of cards of analytical distinctions and concepts. Finally, laypeople must give up the belief in a private and cherished inner world. Our experience no longer remains inside a reassuringly sheltered and private mental domain. Our experience is the world we live in. We are cast outside of our skin, so to speak.

The good news is that neuroscience will be able to reinterpret empirical evidence. Scholars will stop looking for mental properties as opposed to physical properties. Laypeople, too, will benefit from being freed from a claustrophobic, inner neural world.

Consciousness—as is traditionally conceived—cannot be seen, cannot be measured, cannot be observed. We can see the footprints left by people's minds, but we cannot see their minds. Has anyone ever seen a mind? I have never seen my own either! I have seen objects, but I have never seen *an experience*. In fact, has anyone ever seen an experience? I do not think so. Given the common worldview, experience is invisible. Scientists, as well, have never observed anyone's consciousness. They observe other things, such as one's behavior or neural firings. We see vividly colored fMRI images (Functional Magnetic Resonance Imaging) in respected scientific journals. Such things, though, are not one's conscious experience. Has any neuroscientist actually observed an experience, as experience is described by the standard view? Never. At most, they have recorded physical phenomena—say, neural activity in one's fusiform gyrus—that, in standard conditions, co-occur or correlate with experience.[2] As the philosopher Tim Crane observed in 2017, "fMRI technology does not solve the mind–body problem; if anything, it only brings it more clearly into relief." This deafening lack of direct evidence is highly suspicious, to say the least. Why should experience be beyond the natural order? In contrast with this picture, I expect to find experience somewhere within nature, not unlike the way I find electrons, pebbles, and parrots.

In fact, many have observed that our experience is not different from the surrounding world.[3] In fact, it is made of objects, people, cars, buildings, trees, clouds, the sun, and stars. In our life, no difference separates, say, the sun I see in the sky and the sun I experience. Everything that is part of my life is an object. Everything is physical. In turn, physical objects have a causal role and, finally, they are part of something we call the *now*. So, in the physical world made of objects, what is the thing that is identical with my experience? Furthermore, *what is the thing that I am*? What are my physical, temporal, spatial, and causal limits? *Where* and *when* does my experience begin and end? If I were only my brain, how could I experience the world outside my body? How could I experience a world that is distinct from my body? Am I truly inside my body?

To all these questions, the idea of spread mind provides a unified and original solution that outlines *what* and *where* consciousness is. It shows that the experience, say, of the gracefully blossoming willow in front of me is identical with the gracefully blossoming willow. If this view has any merit, it insists that one's consciousness is not inside the brain, but it is identical with the objects surrounding us. One's consciousness is then *outside* one's body. We are the world around us.

Such a solution faces, at least, three sets of objections. The first is represented by the assumption that, by means of a private inner realm, appearance and reality are separate. The second is a bundle of arguments based on misperception, illusion, and hallucination. The third is an oversimplified notion of the physical object. However, I am confident that the following pages will address all such worries. From the onset, I want to stress that spread mind is an empirical hypothesis about the physical underpinnings of consciousness, not simply a conceptual sleight-of-hand. It is a strong physicalist framework aiming to place consciousness in the physical world. Such a hypothesis is capable of generating predictions that are amenable to falsification. Therefore, the theory of spread mind qualifies as a scientific theory about the nature of consciousness.

I challenge the separation between experience and world. Such a gap is based on wrong beliefs about the nature of both objects and experience that have never led to a clear understanding of the nature of consciousness. Conceptual distinctions are valid only if they lead to successful clarifications. Otherwise, they must be questioned. I will show that, whenever it seems that a physical object—say, our beloved red apple—does not match one's experience, in a rather unexpected twist, a physical object is actually *there*. Of course, to take such a step, the notion of "there" requires a revision of current parochial limits. "There" will encompass a spatiotemporally extended world—the world we live in. Once the notion of "there" is correctly conceived, we will be ready to embrace a radical step—namely, that one's experience is not a neural process in the brain but rather it is the external object.[4]

It is worth mentioning a conversation Ludwig Wittgenstein had with a friend. According to philosophical gossip, the philosopher asked a friend, "Why do people always say that it was natural for men to assume that the sun went around the earth rather than that the earth was rotating?" His friend replied, "Well, obviously, because it just looks as if the sun is going around the earth." Wittgenstein rebutted, "Well, what would it look like if it looked like the earth were rotating?" Likewise, if a skeptic thought that placing one's experience in the external object is counterintuitive, I might reply, "Well, what would it look like if it looked like our experience were the external object?" and my guess is that it would not look any different. The spread mind account is consistent with our experience of the world—it conflicts only with entrenched conceptual frameworks.

The idea of spread mind eliminates any need to posit a contrast between appearance and reality. Illusions, dreams, and hallucinations are explained away in terms of physical objects about which we have held wrong beliefs. In order to dream a pink elephant, I must have met, in a way this book will shortly explain, a real pink elephant in my life. A careful review of the evidence for cases in which subjects have allegedly experienced pure mental properties—from Penfield's brain stimulation to supersaturated hues—will reveal that, in fact, such cases have never occurred. I will do my best to address all cases in which, apparently, one perceives something—like a pink elephant—that does not seem to exist. I will show that the difference between appearance and reality has been greatly exaggerated. Hallucinations do not differ from perception in their causal structure but differ to the extent that one cannot interact with their objects. The unexpected conclusion is that *all experience is perception of external objects.* In turn, *perception is identity with such objects.*

The theory of spread mind is based on a hypothesis about how to carve nature so that experience and the world no longer run afoul of each other. Once objects are reconceived in causal and temporal terms, they match so closely with experience that the distinction between consciousness and the world can be forgotten as the relics of a bygone age. One's experience is identical with a part of nature and, in turn, everything in nature—red apples, the experience of red apples, dreams of red apples, and so forth—is physical.[5] It cannot be anything else. The Magritte-esque notion—so perfectly rendered in his 1933 painting *The Human Condition*—that we are acquainted with the physical world simply by means of internal representations, mental states, phenomenal experiences, and qualia is only the offshoot of erroneous interpretations of empirical evidence. Enthralled by the wonders of perception, the modern tradition has been too hasty in drawing conclusions as to the metaphysical difference between experience and the world. The theory of spread mind suggests that the world is not different from what it appears to be. The only required relation is the simplest one—namely, *identity.* The details of how this basic intuition works constitute the bulk of the next chapters. All well-known cases of conscious experience—such as perception, misperception, memory, mental imagery, afterimages, illusion, hallucination, phosphenes, and phantom limbs—will be addressed in terms of identity with physical objects. I will argue that what we call experience is the part of the world that is identical with us, moment by moment. The bottom line is that *my experience of the round red apple placed on the table next to me is one and the same with the round red apple*. No more and no less.

· · ·

I look around. I see a world made of objects and their properties. By and large, if something is physical, it should be locatable in space and time. Everything, since it is a part of nature, must be somewhere and at some time. No exact punctual location is required. A location can be fuzzy, vague, probabilistic, and plagued by uncertainty. Then, in the world I live in and perceive, what and where is my experience? Roughly, we know where our bodies are. Likewise, we know where either the moon or a red apple is. Here, however, I do not refer to my body, but to the thing that my experience is—a thing describable in terms of perceptions, dreams, thoughts, and desires. Can I locate my own experience as I can locate the performance of Vivaldi's *Gloria* I heard at the Boston Symphony Hall on the afternoon of March 12 in 2014? Can I locate my dream of a pink elephant in the same way? Where, when, and what is the stuff my dreams are made of? To this day, both scientists and philosophers falter when confronted with such questions—be it a dream, a hallucination, or a perception. At best, scientifically-minded scholars point at correlates, supervenience bases, representational vehicles, and so forth. The hesitation is embarrassing because if one assumes that experience is a natural phenomenon, it ought to be spatiotemporally located. However, putting the mind inside the brain has not yielded any solid finding because, trivially, nothing inside a brain has the properties of our experience. So, again, *what*, *where*, and *when* is experience?

The hypothesis I am putting forward is that our experience of an object is not where the body is, but rather *where the object we experience is*. More precisely, I suggest that *our experience of an object is the object we experience*. So far, due to various entrenched misconceptions, most researchers have been looking inside the body for consciousness. Only a handful have considered the processes taking place between the body and the world.[6] Here, I dare an even bolder move—*one's experience is the very object one experiences.* Experience is proposed to be a part of the external world. It is not a neural process in the brain. It is not a process enacted between the environment and the brain through the body. *Experience is the external object.* For example, it is the red apple you can grab and eat. It is the cloud that looms above your head. It is the sun that shines in the sky. Thus, your experience is a part of the world. It stands in front of your body. *Your experience is in front of your eyes, rather than behind them.* While your body is here, enveloped by the skin, experience is elsewhere—more precisely experience is there, in front of your eyes—*physically there*—and not behind them. You are there too where your experience is. You are not behind the eyes. Your body is behind them. You are beyond them. The red apple is as physical as the neural activity that takes place in your brain: there is no risk of stepping outside physicalism.

Experience is the thing one looks at. The mind is the world one's eyes look at. Hence, experience is a physical phenomenon, only not a neural process. This hypothesis is admittedly radical, but it is of an empirical nature, since what is in front of the eyes is as physical as what goes on in the brain. The red apple is as physical as a neuron.

If experience is physical, it must not only take place at a given spatial location, but also at a given temporal time. So, *when* does my experience take place? Time may seem less demanding than space, but, as we will see further on, the present we experience is at odds with most notions of time. If we experience objects that are far away in both space and time—e.g., stars—when does our experience take place? What about cases—such as memory or dreams—when we experience past episodes of our lives? The standard conception of the present suggests that everything that matters must be temporally located in some very short and rather ill-defined temporal neighborhood of the current instant.

Such a punctual notion of the present is thorny because—given the temporal extension of physical processes—nothing fits inside such a narrow span. Experience is no exception. The external object, such as the distant rumble of thunder, is allegedly temporally separate from the neural activity going on in the brain. In this regard, the standard view endorses a gap in time as well as in space, not to mention a gap between appearance and reality. In contrast, the theory of spread mind places experience at the time when the object takes place. The time of one's experience is the time of the objects one experiences. Therefore, if I look at a star fifty light years away, my experience is temporally located fifty years ago. More on this later.

At this point, an objection is to be expected. To many readers, a view stating that consciousness resides in objects rather than inside the brain might appear to be a scientific nonstarter. Honestly, though, I do not see any strength in this objection. Objects are a part of the physical world. No additional fancy entity has been invoked—no aura, no ghost, no soul, no spirit, no holistic entity, no invisible/emergent/intrinsic property, no integrated information, no quantum mechanical mystery, and no dual aspect of reality. I defend the view that experience in all its forms—from everyday perception to hallucinations—is out there in the environment, just like streams of water, pebbles, and stars; that experience *is* water, rocks, and stars. The theory of spread mind is an empirical hypothesis as to the physical nature of experience and consciousness. Why should the identity between experience and external objects not be scientifically respectable? Is there anything more physical and concrete than objects? I concede that, at first sight, identifying consciousness with external objects might appear strange, but no *a priori* reason forbids one's experience from being physically located outside one's body. In this regard, the philosopher Daniel Dennett, in his philosophical tale *Where Am I?* (1981), remarked that we have no introspective access to the location where consciousness occurs. The only criterion for choosing the location of experience is finding something that has the same properties of experience. No empirical evidence prevents the experience of the red apple from being located where the red apple is. No fact prevents my consciousness of the red apple from being one and the same with the red apple.

Relocating experience in the world—and therefore "spreading" consciousness in space-time to such unheard-of latitudes—can pay back in terms of simplicity. If experience is one and same with the world, no chasm will open in the fabric of nature. One will dismiss those notions that have never matched with the natural world, such as representations, phenomenal characters, mental properties, and so forth. Consciousness will no longer be an unexpected addition to the physical world. Appearance and reality will be the same thing. Endlessly multiplied intermediate entities and relations will be set aside. Identity will be the fundamental—and only—relation.

The key hypothesis here is that one's consciousness is identical with the very objects one experiences. With a linguistic twist, one might morph William James's "a world of pure experience" into "an experience of pure world." *Consciousness is where and when the physical objects that one experiences take place.* It is not an additional, invisible, and unexpected phenomenon underpinned by neural activity. Experience lies not inside the body, but rather is the world we experience. The mind is spread.

.   .   .

Is identity between experiences and objects a seriously defensible claim? To defend it, consider standard perception. Suppose Emily perceives a red apple. Suppose also that you have not yet signed up to the view that what the world is and how the world appears are two separate ontological aspects. Emily's body B and the red apple O on the table are all we need.

O   B

Thus far, no extra property is needed to model what is going on. Emily's body is numerically different from the red apple. The red apple is on the table. Emily's body is roughly one meter away. These facts do not pose any special problem. However, what is Emily's experience of the red apple? Where and what is Emily's experience of the red apple? If one is an internalist, one is tempted to posit an additional invisible entity—namely, consciousness or experience—because nothing inside Emily's body is like the red apple and consciousness is expected to reproduce the external world. On the other hand, if one is a strict realist,

Emily's experience accesses an external object via some unknown relation. Unfortunately, neither notion of experience makes any sense in physical terms. Both Emily's body and the red apple are physical objects, and physical objects do not partake of other physical objects. The body cannot access external objects. B is B and O is O. The recourse to the notion of person as something additional to one's body is not convincing since, in physical terms, nobody has a clue about what a person is. The appeal to the notion of person is a case of *obscurum per obscurius*—namely, explaining something we do not understand by means of something we understand even less. In the physical world, no fanciful relations between objects have ever been spotted. Objects are just what they are. Red apples are just red apples. Human bodies are just human bodies. Neurons are just neurons. Period.

Are we stuck in a hopeless situation? We will be, if we stubbornly insist on looking for experience only inside Emily's body. The mind-body problem and all of its cognates arise from the assumption that Emily's experience (E) has to be either inside/dependent on/constituted by/supervenient to her body. In fact, once we assume that Emily's experience depends intimately on what is going on inside Emily's body, the experience of the red apple is impossible because Emily's body and the object are distinct and separate. Moreover, they have entirely different properties.

For dualists, the problem is easily solved. Dualists model experience as a sort of ontological joker that plays the role they want it to play. However, the price we pay for such a wild card is high. Experience is conceived of as an additional entity that does not match with the physical world. We are all aware of how thorny the case is. Fortunately, the theory of spread mind suggests a different solution.

For a second, set aside your beliefs that your mind is inside something. Consider the possibility that experience is not located inside the body but rather that it is in the perceived object. Let us be bolder!

Suppose that *experience is the external object itself.* Experience is not intimate with one's body. Experience is intimate with the object. More radically even, experience is the object. *One's experience of an object is the object one experiences.*

$$O = E \qquad B$$

The physical properties of the body and the object are known approximately. The brain is pinkish-gray, gooey, and bloody. The red apple is red, round, and applish. Emily's experience is red, round, and applish. Which entity is more similar to Emily's experience of the red apple? What is round, red, and applish? The brain or the red apple? The answer should be obvious—the red apple. When one experiences the red apple, something has exactly the properties one experiences—namely, the red apple. *E is identical with O.* Experience—i.e., consciousness—does not need to emerge from the physical world, since it already has the properties of the external object, since the two are the same. Conversely, the difference between the brain and one's experience is no longer a reason of concern. Consciousness is not a property of the brain. The object causes an activity in one's body, but it is not that activity. In addition, experience does not need to be connected to the object by means of some vague relation. Consciousness and world are one. E is O.

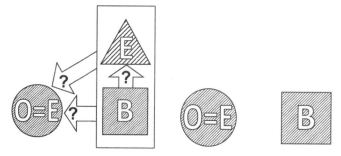

Figure 1. On the left, the traditional view strives to connect object, body and experience with as many problematic relations. On the right, the spread mind.

What roles do body and brain have in singling out objects from the physical continuum—the physical intercourse usually dubbed perception? The traditional account tells us that the body accesses objects through the senses and then concocts mental percepts. According to the view presented here, the body is key for the existence of the object, which remains in the outside world. The body offers the causal circumstances that allow the objects that one experiences to exist. The objects one experiences do not exist autonomously. They are causally relative to our bodies. The objects that compose our world are relative to our bodies. They are relative objects as velocities are relative to a frame of reference. They are neither mind-dependent nor mental, they are physical objects whose existence is causally based on our bodies. The objects we experience are body-dependent. Therefore, the body does not produce an additional entity called experience; rather it is responsible for the existence of the object, which is identical with one's experience. The existence of the object depends causally on the body but it is different from it. O is not B. Neither E is B. Rather, E is O. In such a physicalist landscape, there are only physical objects that are causally related. The body singles out a specific experience by singling out a specific object. If the body were not there, the object one perceives would not exist. Such a claim is not a form of idealism, since no entity is mind dependent. Here I endorse object-dependence, which refers to the fact that each object exists relatively to another object.

To shed a different light on the relation between the object and one's body, consider the analogy of a system composed by a lake and a dam. The water flows in a river. Because of the dam, a lake fills up. Before the dam was built, no lake filled the valley. However, because of the dam, the water formed a lake. The water is not just the lake and the lake is not just water. The lake exists because of the dam. The dam was built because of the river and the possibility of creating a lake. The dam does not create the lake out of thin air, though. If no dam had been erected, no lake would exist. Crucially, the lake is not an emergent property of the dam. If there were no rain, a one mile-high dam would not create a lake autonomously. Rather, the lake takes place because of the dam and the river, and because of several other conditions. Yet, in that valley, the dam has a contingent role in bringing the lake into existence. To extend the analogy, the lake is the object. The river is the incoming flow of events. The terrain is the environment. The water is the stuff the world is made of before one experiences it. The dam is one's brain and body. Where is one's experience? It is the lake. Where is the object? It, too, is the lake. The key idea is that the existence of the object is dependent on the proper causal circumstances. Given a certain environment, the body provides and completes such critical circumstances.

A few preliminary remarks about the distinction between the phenomenal and the physical are necessary. It is commonly assumed that phenomenal properties appear to have "quality" or "phenomenal character," while physical properties lack these attributes. Since Galileo's seminal work *The Assayer* (1623), in which the Italian scholar set subjective qualities and objective quantities aside, such an ontological apartheid has been nothing short of a dogma. Yet, is such a gap really so obvious? Isn't such a gap so ripe with riddles and mysteries that one should be cautious? For instance, the gap between phenomenal and physical boundaries has endorsed the bizarre idea that the physical world is invisible and unperceivable and that physical properties are disguised by phenomenal properties. How can brains, which are physical, trigger *unphysical phenomenal appearances?* So far, nobody knows how.

The theory of spread mind adopts a different strategy by asserting that phenomenal and physical properties are the same. *We perceive the world as it is because we are the world we perceive.* The properties of our experience are the properties of the physical world we live in. To be clear, I do not claim that physical objects transmogrify into phenomenal ones. Nor do I claim that physical objects host phenomenal qualities. No panpsychism is advocated here. My claim is that there are only physical objects and physical properties: that one's experience is identical with them. In this way, there is no tension between unperceivable-yet-physical objects and perceivable-yet-immaterial appearances. There are just relative objects all the way down, say red apples, which have been split erroneously into two fictitious concepts: the physical object and one's experience of it. In the theory of spread mind, the step from physical to phenomenal is not a step at all: no emergent processes transmogrify the physical into the phenomenal. When I perceive a red apple, if my experience is one and the same with the apple, no mystery remains.[7]

In fact, the identity between experience and external objects appears so obvious that I feel the need to outline a tempting hypothesis as to why so many sophisticated scholars have insisted on placing experience inside the body. The main reason, which points to the usual culprit, is likely represented by cases of misperception (perceptual error, illusions, and hallucinations). Moreover, human beings have an irresistible tendency to place experience inside the body because of the attraction exerted by their "center of perceptual gravity"—to borrow Dennett's felicitous expression. Moreover, one must preserve the body from harmful circumstances in order to stay alive. Not to speak of the social, aesthetic, and legal pressures to identify a person with one's body. Finally, a vestigial anthropocentrism resists the idea of considering experience and things—i.e., man and nature—on equal footing. Taken together, these factors have conspired to keep experience and objects apart.

Is asserting the identity between experience and physical objects too naïve? I do not think so, since any physicalist theory aiming to explain a phenomenon must put forward, at the end of the day, a hypothesis about what the phenomenon is identical to. Temperature is average kinetic molecular energy; evolution *is* variation, selection, and transmission; pneumonia is a proliferation of *Streptococcus pneumoniae* in the lungs. And so forth. *The physical must be explained by the physical only.* In this respect, the theory of spread mind is not any different: it puts forward a hypothesis as to the physical nature of experience that will stand or fall based on empirical evidence. The notion that experience is identical with external objects sounds strange at first, but it does not offer any reason for outrage. If experience is physical, it must be somewhere, as all that is physical. Since Galileo, the traditional reply has been that consciousness is somewhere inside the body rather than in the external world. The theory of spread mind suggests a different location: the object itself. Being a red apple is as good a hypothesis as experience being a neural process. The two hypotheses are both empirical claims and must be judged by the available evidence.

Moreover, the claim that experience is located inside the body—or that experience is an emergent property of something located inside the body—is committed to the idea that experience is different from the thing inside which it is located. This is a dualist attitude. In the physical world, experience ought to be identical with something physical. However, experience cannot be inside the body because, inside the body, nothing has the properties of one's experience. The closest thing to our experience is the external object. Once the object is conceived, no difference divides our experience of objects from the objects we experience.

.   .   .

If my experience of an object is the very object I experience, what and where am I? I am not where my body is, although my body is necessary for my existence. While I cannot do justice to the notion of the self here, relocating experience also relocates the subject. However, for the sake of argument, as a provisional notion, I propose that, at any one moment, one is identical with one's consciousness—i.e., with one's experience.

For the sake of simplicity, consider again the red apple. As I write this, I am sitting in the central aisle of the Bates Hall of the Boston Public Library. I see my hands, I see the laptop on the table, and guess what, I see a red apple. In the distance, I see a long line of similar desks sparsely occupied by students. I hear a background buzz coming from the many people loitering, reading, and writing in the vast hall. Consider the traditional visual field. According to the theory of spread mind, such a "phenomenal visual field" is not a mental projection—it is a portion of the Boston library that my body carves out by means of the particular causal intercourse usually called perception. My experience is not a mental image but a portion of the hall. I do not see an image. I see the Bates Hall or a part of it. Seeing is being. I am the Bates Hall. More precisely, a portion of the Bates Hall is a portion of my own experience. This claim might sound bizarre, since it is often stated that one's view of the world and the world one beholds are different both numerically and in kind. Yet, my experience and a part of the Bates Hall are much more similar to each other than my experience and my brain activity. Why shouldn't I be made of the very world that surrounds my body, that exists relative to the causal circumstances offered by my body?

Emily—who is now sitting next to me—does not look at the red apple only. Her perception (and therefore her consciousness) includes my laptop, the table, various lamps and a long list of other objects and bodies that, in this moment, make up both Emily's experience and a relevant portion of the Bates Hall. We do not need to introduce an inner world nor do we need to contrast the physical hall with a mental one to arrive at an explanation of my experience.

Emily's experience, too, is identical with a subset of Bates Hall, which is singled out by causal processes that her body makes possible. Her experience is a subset of what takes place inside the Bates Hall rather than a subset of what goes on inside Emily's head. Some of the objects in the library are part of Emily's experience, others are not. However, we do not need to double the ontology of what is going on inside Bates Hall. Emily's consciousness is made up of a subset of the available things inside this library.

Emily's consciousness is both physical and observable—no mysterious, hidden, private mental states are invoked. It is there, all around her body. Experience is not an inner world. When I say to Emily, "Look! That is the red apple I see," unless Emily is somehow causally impaired with respect to the apple—for instance, she might be congenitally blind or blindfolded, or the apple might be occluded—Emily will see the same apple that I see right now. To a large extent, the same apple will be part both of Emily's experience and mine. Seeing the apple is neither the concoction of a mental image nor the instantiation of an esoteric relation between an object and an enigmatic subject. *Perceiving an object is being one with such an object.*

The fact that the red apple is observable does not mean it has to be easily observable by everyone. In practice though, the case is no reason for concern. When two or more people are looking at the same object, no one is really wondering about whether they see the same thing. When Emily and I look at the same red apple on the table—and as long as we both know where the other one is looking—we are confident what the other sees. A red apple. The same red apple! The traditional view that Emily perceives Emily's private mental version of the apple and I perceive my own—both different from the actual apple—is preposterous. We perceive the same apple. The same red apple is part both of the set of objects that is my experience and of the set of objects that is Emily's. Hence, not only are our experiences alike and share the same properties, but they are made of the same red apple—in fact, they *are* the same apple. When we talk about what we see, Emily and I understand each other, because, to some extent, the same object makes both of us. To communicate is to overlap—it is not an exchange of bits devoid of meaning and content. In this case, the red apple placed on the table, in the central part of the left aisle of the Bates Hall in the Boston library partakes both of her experience and of mine. Emily's experience and mine overlap partially. Unsurprisingly, I am made of other objects and events. Likewise, Emily too is made of other objects and events—some of which are undoubtedly different from mine. Emily's mind and my own do not totally overlap (Figure 2).

Figure 2. Riccardo's experience (left) and Emily's experience (right) are two collections of objects. The red apple is shared between them.

However, if Emily's experience and mine were made of exactly the same pieces, Emily's mind and my own would be the same mind. In fact, it would be difficult to argue against such an identity. In practice, such a complete overlapping never occurs. Two individuals never completely overlap, though kindred spirits get close.

Once again, the identity between a set of objects and one's consciousness might seem bizarre. However, why should any subset of the physical world be any worse than any other one? Why should objects be worse than neurons? Objects look much more promising too. The traditional hypothesis that one's experience of a red apple is identical with a brain process is not more scientifically respectable than suggesting that one's experience is identical with the red apple. Brains and apples are both physical entities. Actually, brains score much worse since they do not have any of the properties of experience while apples have all the properties that one needs. The point I wish to stress is that the explanatory structure of the theory of spread mind is akin to other, apparently more established, views. Here I put forward a hypothesis of empirical nature that is extremely parsimonious, ontologically. It is a strictly physicalist identity theory analogous to mind-brain identity theories. Yet, the suggested physical phenomenon is not the brain but the external object.

· · ·

One of the main reasons why the solution presented has never been thoroughly taken into consideration is the widespread over-simplistic notion of the object. Objects are often described by philosophers as atemporal discrete individuals with instantiating properties and by scientists as aggregates of molecules held together by force fields. Both views are a far cry from our experience of the world. Here, a causal view of objects capable of combining these two views is sketched out. Actually, the theory of spread mind is as much a theory of the object as it is a theory of the mind. In fact, the key insight is that the mind is a set of objects. Most of this book is an effort to show what objects—and thus experiences—are.

The existence of a physical object for every experience is key to overcoming the gap between appearance and reality. Each experience has to be identical with a physical object. Yet, is such a claim feasible? Can we locate a physical object, like an apple or a pebble, for every experience that one has, including illusions, dreams, and hallucinations? Such a claim is, at once, the main prediction of the spread mind and its founding principle.

A strong tradition has nurtured the popular-yet-over simplistic notion of the physical object as something we cannot perceive since, by definition, it transcends our experience. Such an object is customarily described by means of quantitative measurements and mathematical models. In our life, though, we never meet it. It is an abstract wireframe. It is a mathematical ghost, but a ghost nonetheless. Such an object is invisible to the senses and yet we experience it through our senses. Unfortunately, it remains, etymologically and substantially, *ab*-solute from experience.

Yet, when I eat a red apple, I eat a real object. When I see or touch it, I see or touch a real object. When I represent a red apple as a system of mathematical equations, I devise an abstract model that, in many circumstances, allows me to deal with the real object but it is not a real object in itself. The mathematical apple, which is an unperceived abstract notion, is only a conceptual tool. It is similar to notions such as centers of mass—useful abstractions but not physical entities. Notwithstanding these concerns, many are inclined to accept a world populated of invisible properties known by means of purely mental appearances. The extreme case is hallucination, in which—allegedly— the hallucinated object does not even exist. Such a view suffers from circularity. Worryingly, the view assumes what it should prove, namely that what we perceive is not the physical object. It is easy to see that such a notion of objects restates the premise that appearance and reality are distinct. In fact, apart from the various arguments from misperception, neither *a priori* arguments nor empirical evidence back up the difference between experience and the external object. In our lives, everything we experience is real.

I do not want to beat around the bush. An object is something that takes place. Before it takes place, something else, which is different from nothing, is not yet the object. The traditional view assumes that objects lie idly somewhere waiting to be perceived. Perception is taken to be subjective and fleeting while the object is assumed to be stable and durable. Hence, the mind is held responsible for the many different ways in which an object can appear—blurred, in focus, close, far, tactile, visual, and so forth. In contrast, I propose a radical solution—instead of an object being perceived in multiple ways, multiple objects take place. Such objects are causally originated thanks to the beholder's body. Similarly, an object that perdures is a collection of objects that are alike. This is no more mysterious than a shelf filled with nearly identical soda cans. They all look the same, yet they are numerically different. Similarly, whenever I look at the table where certain molecules and cells are arranged in a given way, an apple occurs. When I look again, a new apple occurs. Actually, every time I look back at the table, a new apple occurs. But not a *mental* visual apple, rather a *physical* visual apple on the table. One concludes—and from a practical perspective it is the sensible thing to do—that only one apple exists. Every time I see an apple, I do not see a mental apple, or a phenomenal apple, or a Berkeleyan apple that my mind creates within a mental domain.[8] Rather, I see a physical red apple and such an apple is my own experience of the apple. The apple that takes place every time I look at the table is a physical apple identical with what people usually believe a respectable apple should be. My body, too, is an object. In this case, my body provides the missing pieces necessary for the stuff on the table to take place as the apple, which is my experience. When my body is in the right causal connection, the apple takes place. As it happens, I cannot avoid seeing the apple if my body is in a certain place. Every time my body is there, the apple takes place: I cannot avoid my body bringing into existence the apple[9] unless I destroy either my body or the stuff on the table, or both. Only one apple exists. Due to nomological constraints—i.e., the way in which our universe is built—the apple can be eaten only once but can be seen or touched many times. The bottom line is that no ontological gap separates the appearance of an apple from the apple.

We have been lulled into Galileo's dream for way too long. We have come to believe that the world is populated by mathematical entities that do not depend on our presence. In fact, if our experience were the result of immaterial souls, it might be a reasonable hypothesis. Yet, if our presence entails also the presence of our bodies, we must wonder whether our bodies, which are physical entities, make a difference in the physical manifold in which they are embedded. Indeed, I argue that our bodies enable processes that change the ontology of the world. Our bodies bring into existence the physical objects with which our experience is identical. We are our experience. *We are not our bodies.* We experience the part of nature whose existence is a causal by-product of our bodies. We are one and the same with those objects that take place thanks to our bodies. Of course, such objects are not inside our bodies. Bodies are neither necessary nor sufficient causal conditions for such objects to take place. They are, however, contingent causal circumstances. It is enough that, like Donald Davidson's striking of the match, the existence of a certain body in a certain situation causes a certain object to take place.

A better account of the physical object is paramount because the mind-body problem stemmed from an over-simplistic notion of the object. Since the scientific description of objects does not match our experience of the world, a complementary notion of mental domain or of phenomenal experience has been devised—the left-over strategy.[10] The traditional notion of the mind is a theoretical by-product of the acceptance of an a-causal and a-temporal notion of the external object. The mind is a sort of conceptual crutch devised to safeguard a simplistic object. It fills the gap between the object imposed by scholars and the real object that one experiences. On this score, the mind-body problem is akin to the ancient geocentrism when astronomers invented fictitious trajectories for the planet—the infamous epicycles—only to avoid calling into question the central position of the earth. It is a recurrent scientific fallacy. Scholars do not question a wrong and fundamental assumption and seek convoluted explanations that promise to save their beliefs. To dispatch vestigial forms of dualism, I revise the traditional notion of absolute object and consider a new notion of relative and causally active object. Such a "better" notion of the object does not exert any strain either on experience or on empirical evidence.

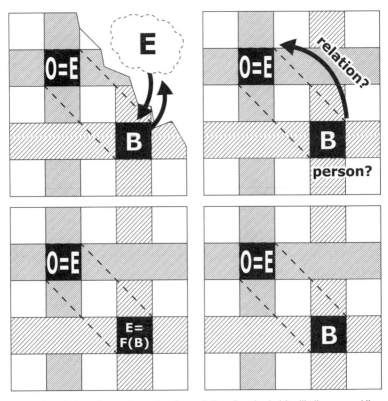

Figure 3. Models of experience. Dualism, relationalism, brain-identity theory, and the spread mind.

The traditional notion of objects runs afoul of our experience of reality (Figure 3). The insistence on placing consciousness and experience inside the body has led to countless puzzles and mysteries. The body is physically separate from the world one ought to experience. However, nobody experiences the body. Nobody experiences the brain. Nobody has ever felt like being a bunch of neurons. The theory of spread mind solves the issue by placing one's experience in the object rather than in the body (Figure 3, bottom right).

So far, the punchline is that objects have owed their existence to causal processes—objects take place as causes of effects. We are inclined to think that an object appears in different ways. This is a delusion. In reality, multiple objects occur and we categorize them together as one. However, relative objects do not require any special ontology—they take place whenever a human body is in the right causal connection with the environment. Each causal occurrence singles out a different object—a visual object, a tactile object, an auditory object, a multimodal object, and so on. Apart from the emphasis on causal genesis, the causal object, outlined here, is just an object—namely, a physical cause in one's environment. The object is a physical cause that produces physical effects. One's experience is made of such objects. Reality is not divided into what exists and what appears. This book is devoted to clarifying under which conditions some of these causal occurrences and their objects form a unity we call mind or consciousness. It is worth stressing that, so far, we have not recruited any psychological notion. We will not. Objects are just physical entities and experience is identical with them. *We are the world and the world is us—everything is physical.*

·   ·   ·

Experience and objects have been kept apart by yet another prejudice: time is not an essential aspect of objects. In contrast, experience seems intrinsically connected to time, to change, and to becoming. While a timeless experience is inconceivable, a timeless apple seems feasible. Yet, is that the case? Putting our prejudices aside, who has ever seen an apple outside time? Who has ever seen an apple that was not (albeit slowly) growing, getting ripe, and eventually rotting? Do actual objects not partake of the causal flow of time as much as experience does? Once again, a simplistic view of the object has so far prevented objects and experience from revealing their identity and has encouraged the invention of an additional mental domain. In contrast, the theory of spread mind puts forward a temporal and causal version of the object that endorses a unified description of reality and experience.

Consider our beloved red apple. Usually, when one describes a fruit, one feels no need to refer to any temporal aspect. The apple is, first and foremost, an object with properties such as color, smell, taste, weight, and so forth—all physical properties. Likewise, the standard description of objects gladly leaves most temporal considerations aside. Objects are conceived of as temporally invariant entities, at least within a reasonable time span. Change, time, and becoming are taken to be nuisances one can largely ignore. Here, I do not refer to the obvious fact that even the most static and everlasting porphyry statue is slowly degrading and changing. I refer to the fact that the objects we perceive cannot be atemporal objects, because they are embedded in the same causal processes that produce our experience. In other words, the object we perceive is the one that takes place when our body is causally tangled with it. I cannot perceive what is not taking place because it would be *causally out of reach*.

One can rebuke that the apple, after all, does not do a lot—it just lies idly on the table. Such a description greatly underestimates what the apple does. Whenever we look at the apple, the apple is reflecting and bouncing light rays. To be visible, the apple does substantial causal work—the apple diffuses light rays. If an apple were lazy with respect to light rays, it would be invisible. Likewise, we smell the apple because the apple is releasing a certain number of chemicals that diffuse into the air. When I grab the apple and then I feel its texture and weight, the object exerts various resistances on my fingers, as well as pulling other masses through gravity. If the apple was causally inert, the apple would be unperceivable.

Furthermore, we live in the physical world, which is a temporal world. Causal processes are embedded in temporally extended processes. The object one experiences is a temporal object. Whenever one looks at the red apple, the red apple takes place and is part of one's experience. Consider watching a movie of a red apple, a very boring movie of a motionless red apple on a table. Any time you look at the silver screen, you see the projection resulting from a different film frame dropping exactly in front of the light beam because of the wonderful precision of the intermittent sprocket. Any time you cast a glance, you see a different film frame, namely a different physical object. Yet your visual experience is indistinguishable from that of a juicy red apple. When you look at the juicy red apple, a recurring physical process from the apple to your eyes repeatedly takes place. Many times per second, a different train of light rays travels from the apple to your retina. What your experience is made of—the red apple you perceive—is a series of identical red apples taking place. The red apple, which is also one and the same with your experience, is taking place right now.

The film is an intuitive demonstration of the fact that we do not perceive a static reality, but instead a temporally recurring world, albeit one that is very coherent and that conveys a substantial feeling of *stillness*. When we watch the movie of a still object, we perceive a still object that does not change. Yet, we know that, at any moment, we perceive a distinct film frame, which is a different object. Standard perception is akin to watching a movie, only we do not perceive a mental movie, but a sequence of identical apples that we lump conveniently together.

It is worth referring to Alfred North Whitehead's notion of Misplaced Concreteness—i.e., the error of mistaking an abstract notion for the concrete. We have historically developed an abstract notion of the object and we mistake it for the real object. In practice, we have multiple objects, such as the apple we see, and a useful abstract notion—the ideal apple. We take the abstract notion for the real thing and we downgrade the actual apples to mere appearances. Then, since an abstract apple is beyond everyone's grasp, a gap between appearance and reality is proclaimed.

A different approach is defended here. The natural world is made of objects taking place in space and time. Experience too takes place in space and time. Once the common causal and temporal nature of both objects and experience is revealed, the identity between experiences and objects is manifest. Objects and experiences are made of temporally located contingent causes. The claim is that our experience and the world we experience are the same set of causally singled out objects: *I am world and the world is I.* The apple I see takes place because of the causal processes between the stuff on the table and my body. A mental apple is no longer needed once we have a causal and temporal notion of the object. One apple is enough to be both on the table and in my experience. We can set aside the invisible apple nobody has ever seen.

Human consciousness has historically been described as a window on the world as in Magritte's archetypal masterpiece *The Human Con-*

dition (1933). However, a window is not the right metaphor. On the contrary, it is a deeply misleading analogy. Rather, one's experience is the object *outside* the window. Inside the window, there are only the causal conditions that enable the external object to exist. In fact, there is not even a glass but a frame that singles out the object in the world. The frame allows the object to take place. Such an object is spread in space and time. Magritte's window is blind, with respect to anything directed inward. *Our experience is outside the fence of the senses, not behind it.* In a sense, Magritte's room is empty. Our body is the frame. It is the causal proxy of spatiotemporally spread worlds made of objects and events. Our minds are those worlds. Consciousness does not take place inside our heads: *one's mind is the set of physical objects one's body brings into existence.* We are neither inside the window nor identical with the frame. We are what the frame carves out, which is outside the window (and outside the body).

.   .   .

An all-too-easy objection is the alleged difference between what the world is and how the world appears to be. Is such discrepancy not taken to be an established fact? One might be tempted to reply with a question. How do we *expect* the world to be? How can we know whether the world is different from how it appears to be? Is it *really* possible that the world appears different from what it is? Is it not suspicious that we inhabit a world whose real properties are hidden?

Let us ponder a bit more. The separation between appearance and reality has long been cherished as the hallmark of a fundamental structure of reality. Yet the notion that something looks different from what it is entails bizarre ontological commitments. For instance, it suggests that there is some sort of duality in the world. Every entity is allegedly masked by a look or by an appearance—vaguely reminiscent of medieval notions such as *eidolon* or *species*—that allow entities to manifest themselves.[11] Such an appearance is not a thing in itself, because if it were, it would not be an appearance but instead another thing and thus it could not be seen. It would require a further appearance and so on, *ad infinitum.* In addition, the distinction between appearance and reality suggests that *what appears is not real and what is real does not appear.* Hence, what appears cannot make any difference to what happens, and what happens cannot appear. These bizarre ontological side effects ought to raise serious doubts about the soundness of the separation between appearance and reality. It is as though appearances were introduced as the leftovers of a simplistic notion of the physical world. As such, like all ontological leftovers, the notion of appearance has ended up where all unfitting entities do, in the "dustbin of the mind."[12]

The "leftover-appearance-model" justified, at a certain stage of development in Western thought, the invention of the psychological circus populated by dubious phenomenal entities such as phenomenal character, content, quality, sense data, qualia, and so forth. Etymologically, "phenomenal" means something that has the nature of appearing, while "physical" means something natural and therefore real. Thus, phenomenal properties ought to be what appear, while physical properties ought to be what exist. *Such a separation irredeemably condemns phenomenal properties to be epiphenomenal and physical properties to be un-experienced.* Yet, no one has ever seen a phenomenal property isolated from a physical property. Vice-versa, we have neither *experience of* nor *knowledge about* physical properties that do not influence experience, either directly or indirectly.

If one dismissed, either empirically or theoretically, all cases of misperception, appearance and reality would no longer split: the red of a tomato would not depart from the red one perceives. The theory of spread mind proposes that no difference looms between what a thing is and what a thing looks like. We no longer need to see looks or appearances rather than what the world is. We see actual objects rather than appearances. We see them because we are them. I would look younger, if I dyed my hair, but this does not imply that an appearance of my body occurs as something different from what my body is. If I dyed my hair, my hair would be physically different. It would not just look darker: it would be darker. The point should be clear enough. Nothing appears different from what it actually is. Everything appears as it is. The difference between appearance and reality is conceptual. It has never been ontological. Appearance is reality.

Undeniably, one can mistake one thing for another. It is a familiar situation. Emily picks a poisonous mushroom and mistakes it for an edible one. I mistook a chipmunk for a squirrel. Once a policeman mistook Bob Dylan for a beggar. Marco Polo mistook a rhino for a unicorn. And so forth. Yet, while one can use the expression colloquially, things do not "look like" anything other than what they are. I can have wrong beliefs about what things are expected to look like. But things always look as they are. We mistook misbeliefs for misperception. Physical entities can be similar, in some respects, to confuse beholders. For instance, one night, Bob Dylan had something in common with beggars. Chipmunks and squirrels are both small, four-legged, and furry. Edible and poisonous mushrooms differ only in minute details. Both rhinos and unicorns have one central horn and are four-legged. It is likely that something might be mistaken for something else. However, these mistakes are the result of the observer's shortcomings, rather than the expression of a deep ontological gap. The notion of appearance as something different from reality has always encouraged people to believe that such a metaphysical chasm exists. In contrast, whenever one meets an object, the object looks exactly like itself. Everything is and looks like what it is. Period.

Perception and resemblance have an empirical common ground. Two objects are taken to resemble each other when the subset of properties we perceive is the same—e.g., the horn is shared by both rhinos and unicorns. Two objects look the same when, to a certain extent, they are the same as the philosopher Edwin Bissell Holt stated repeatedly in 1914.

Consider two distinct objects, a red sphere and a blue sphere of equal size. If one perceived only shapes, they would look alike. They would both look like a sphere. On the other hand, if one perceived only colors, they would look different. Likewise, if one were blind, a good portrait of Donald Trump would not resemble the President of the United States. To a certain extent, our notion of similarity is parasitical on our perceptual shortcomings. Likewise, the fact that Daniel Craig's wax statue looks like Daniel Craig shows that Daniel and his statue are alike in shape and that our perceptual system is shape-oriented. Yet, the wax statue does not look different from what it is: some features of the statue are shared by Daniel's body. Daniel's wax statue and Daniel are alike insofar as their observer is interested only in what they have in common. Daniel and Daniel's statue do not look any different from what they are—they *are* the same in some respect. A blind dog would not find any resemblance between Daniel and Daniel's statue.

Nevertheless, common properties do not require the notion of appearing or looking like something else. Two people can have the same hair color and haircut. They can also have the same height and wear the same uniform. Thus, they are identical as long as one considers just height, clothes, hair color, and haircut. It is not that they are similar in their look or in their appearance. They are similar in their actual physical properties. Of course, nothing prevents us from using the locution "they have a similar look." Such an expression refers either to the fact that if one looked at them from a distance, one might not distinguish them, or to the fact that if one were interested in their haircut alone, they would be identical. However, it is just a figure of speech.

• • •

Let us consider a wonderful possibility—the existence of a place where no misperception ever occurs: Shangri-La.[13] Shangri-La submits to the same physical laws as does our world. Neither metaphysical nor nomological allowances separate our world from such a wondrous place. All the peculiarities of Shangri-La, however improbable they might seem, are entirely of a contingent nature. In fact, Shangri-La might really exist in some remote location sheltered by impenetrable mountains.

In Shangri-La, due to a series of fortuitous circumstances that led to genetic mutations, people do not dream. Moreover, they have never experienced any perceptual illusion, nor have they ever been affected by synesthesia, migraines, or hallucinations. Shangri-La natives have eyeballs that cannot be displaced because, due to another minor genetic mutation, their eyes are powerfully held in place by exceptionally strong muscle fibers. In accordance with their religion, people neither draw nor paint. They have never seen a Müller-Lyer illusion. They do not have patterns on their walls, nor do they like to spin disks—so no accidental Benham's top ever existed. Moreover, thanks to a surprising genetic leap, their retina inverted the directions of cones and rods so that blood vessels and nerves are on the external side of the eye; Shangri-La natives do not have any blind spot. Finally, due to the lack of required patterns in their environment and to the height of surrounding mountains preventing direct sunlight, natives have never experienced afterimages of any kind. In sum, aside from a few genetic mutations and a few environmental oddities, the people in Shangri-La are not different from us: they are still *Homo sapiens.*

Life in Shangri-La may seem dull to us, but its inhabitants carry on a very serene existence. Notwithstanding, or perhaps even because of, such an uneventful life, natives indulge in philosophical discussions. Since they have never had any experience that was not readily reducible to a physical object, they have not developed any argument from misperception. In Shangri-La, the dominant theory of perception is straightforward—perceiving an object is being that object. In fact, in Shangri-La, natives do not use the adjective *physical,* since they have never had any reason to distinguish physical from mental entities. Whenever Shangri-La natives perceive a red apple, there is a red apple. Whenever they perceive red, there is something red. Everything one experiences exists. Misperception never occurs. In such circumstances, they have no reason to split experience from world, appearance from reality, or subject from object.

In Shangri-La, Cartesian skeptical doubt would be akin to a scientist who would suggest a duplication of all measured quantities, claiming that for every physical property, we must also consider a phantasmagoric property. Phantasmagoric properties are invisible, hidden, private, unobservable, and epiphenomenal. Such a bizarre scientist would claim that, by sheer idle reflection, he has found out that the existence of a world akin to our own but devoid of phantasmagoric properties is conceivable. He would claim we have no way to tell whether phantasmagoric properties are instantiated in Shangri-La. He would dub the nature of phantasmagoric properties the "Hard Problem" and would suggest a dual aspect of reality.

Shangri-La philosophers and neuroscientists have not developed any notion of inner mental worlds. They have no need to distinguish mental from physical properties. They would not understand the need for supervenient relations linking levels of reality. Only reality exists. Shangri-La scholars do not need to complicate their ontology with unnecessary, intermediate, and troublesome duplications of entities—i.e., appearance vs. reality, representations vs. objects, vehicle vs. content, subject vs. object, and mind vs. world. In Shangri-La, seeing is being.

The theory of spread mind will show that we, too, live in Shangri-La. To support such a view, I will show that, as it happens in Shangri-La, for each experience—be it a dream, a hallucination, an illusion, or anything you like—a physical object/event/property exists. Based on empirical evidence, I will make the case that the arguments from misperception are in fact unusual instances of perception biased by wrong beliefs. Thus, *perception is identity with the physical object*. Eventually, we will get back to Shangri-La and find out that we have never left it. We have always lived in Shangri-La. In fact, we live in Shangri-La right now!

2.

The Spread Object

> GALILEO: One of the main reasons for the poverty
> of science is that it is supposed to be so rich.
>
> —BERTOLT BRECHT, 1945

F EXPERIENCES ARE objects, understanding the true nature of familiar everyday objects is key. What can we say about their nature? It might be that, because of historical factors, we have so far been misled by an over-simplistic notion of objects that has hampered our understanding of both nature and us.

Like an arch that bridges a crevasse, a theory of mind must connect world and experience. So far, many authors have focused their efforts on the side of the mind, naively leaving the side of the object untouched—the mind is expected to make most of the effort. I will argue that this is the offshoot of an over-simplistic notion of everyday objects—naïve materialism that fits neither with empirical evidence nor with our experience. The classic object sits motionless, like a princess, waiting to be unveiled and revealed. The existence and nature of objects is assumed rather than explained.[14] Understanding the nature of objects is essential if we want to comprehend our experience of the world. Without questioning fundamental physics, it is worth rethinking the notion of the physical target of experience, popularly known as the *object*. This is all the more compelling because the object I see, touch, and smell, but also grasp, push, pull, throw, and smash, is very different from the classic notion of object.

For the sake of discussion, for the time being, let's keep separate the problem of what it is to have an experience of a red apple (the problem of *perception*) from the problem of explaining how sometimes I can have an experience of a red apple when no red apple seems to be around (the problem of *misperception*). To keep the two problems conceptually divided, I ask you to adopt the Shangri-La scenario, at least for now. We shall only consider cases in which whenever one perceives *a red apple*, lo and behold, a red apple is there.

In Shangri-La, objects and experiences take place together—whenever you see an apple, the apple is there. My claim is that once we upgrade our notion of the physical object (from the Galilean to the spread object), the thing I perceive is not different from my experience of it and thus that experience and objects are numerically the same.

• • •

As of Galileo's time, modern science introduced a simplified notion of physical reality that has proved very efficient in tackling a huge number of questions. Such a simplified notion spawned a correspondingly simplified notion of the object I hereby dub the *Galilean object*[15] because it is reminiscent of the view expressed by Galileo in his seminal work *The Assayer*. The object is modeled as an entity that is absolute,[16] atemporal, and endowed exclusively with quantitative properties utterly different from those one experiences. In Galileo's words (italics mine):

> To excite in us tastes, odors, and sounds I believe that nothing is required *in external bodies except shapes, numbers, and slow or rapid movements*. I think that if ears, tongues, and noses were removed, *shapes and numbers and motions would remain*, but not odors or tastes or sounds. The latter, I believe, are nothing more than names when separated from living beings.

All physical entities and objects received this treatment. For instance, consider light. A popular yet inaccurate model holds that light is defined by its frequency,[17] which is only a number (a quantity), while the perceived color is eventually added by beholders.[18] Sadly, like platonic ideas, the Galilean object suffers from a very aloof attitude...it cannot be experienced directly since experience is assumed to be inside *the living being* and thus it is not in direct contact with any object. The traditional notion of the Galilean object is an example of what Whitehead considered the fallacy of misplaced concreteness (or reification)—namely, to use his formulation in *Science and the Modern World*, "To regard or treat (an abstraction) as if it had concrete or material existence."

The Galilean object has been one of the most successful abstractions ever. Yet, its success has hampered us from understanding our place in nature. In fact, as a result of Galileo, a well-established tradition—both in philosophy and in science—has stubbornly modeled the object in terms of something substantially different from experience, as a Platonic ideal mathematized object that cannot be experienced. The object has been pushed into a noumenal dimension—i.e., it cannot be experienced by us. The objects that science describes by means of mathematics cannot be seen by us; they are only abstractions.

In *Against Method*, the philosopher Paul Feyerabend noted that Galileo introduced a kind of Platonic experience made only of abstract data and numbers. The world we experience around the clock, made of blue skies and red apples, was considered *subjective* and left to psychologists and philosophers. As C. D. Broad put it in 1923, "the philosopher becomes the residuary legatee of all those aspects of reality which the physicist has decided to leave out of account." The layman too was convinced to trust science and to accept a sort of continuous revelation of what reality should be thanks to the superior authority of scientists. Yet, these aspects of reality are the very properties of the world we live in and we experience. They are the true reality. The red apple I see is the real apple.

The alleged difference between the colorful object one sees and the abstract entity scientists measure has found expression and indirect confirmation in the traditional gap between disciplines. Such a separation has flattered people's wish for an immaterial mind and has secured the authority of scientists in stating the ontology of the world. On the one hand, the physical sciences: and on the other, psychology and philosophy.[19] But the physical object—the real one and not some abstract model of it—already has the resources to be one's experience. The real object, which is both the apple on the table and my experience of the apple, can be neither the Galilean object nor the scientific abstraction. The object we experience is identical with the object we encounter in our life. The physical object is your apple, not the lab's version of it.

The Galilean object is, where there are no minds around, absolutely autonomous and silent. It is conceived of as phenomenologically dark. The existence of such an object is absolute in the sense that it does not depend on anything but itself. If the rest of the universe vanished, such an object would not disappear. It would only be cut off from the rest of the universe. It is potentially eternal since it might continue to exist forever. It is atemporal in the sense that time is not intrinsic to its existence. If a god froze time, such an object would not change. The traditional object does not need time. It dwells in only an extrinsic relation with time; an ephemeral uncommitting affair, so to speak. In the standard account, the properties of an object can be completely assessed without any reference to time—they are size, weight, color, length, width, height, and the like.

Galileo and science asked us to believe in an invisible object that nobody can perceive. Such an epistemic move assigned to elite savants the task of defining the nature of reality. The real object, made of colors, shape, texture, taste was downgraded to a mere mental appearance—"nothing more than names when separated from living beings." Not for nothing was Galileo's slogan "doing violence to the senses." From a sociological perspective, it is interesting to see the enthusiasm with which each subsequent generation of scientists has tried to undermine the very root of our experience of the world. Recent attempts to demonstrate that we are unable, not only of perceiving the external world, but even our own experience, are just the most up-to-date examples of such an attitude.[20]

.   .   .

The objects we deal with are relative, active, causal, and temporally constituted. They have properties that are identical with what we experience. The object we experience—which is identical with our experience—is temporally located. We do not perceive a red apple. We perceive a red apple at a certain time, because of a process that, due to nomological constraints, singles out the very apple we perceive. The object we perceive takes place in time. Furthermore, such an object is active since it exerts its influence on our bodies. *The object we perceive is singled out by the characteristics of our body.* The objects we perceive exist relatively to the causal characteristics offered by our bodies. Different bodies pick different objects out of the same physical continuum. The objects we are surrounded by are not aloof entities hiding in some ontological fold of reality. They are good-humored vocal extrovert entities we cannot but meet. They present themselves naturally and without any reserve. They are naked in front of us.

What is an object then? When I look at the red apple, what is the thing that presents itself? The object is *causal, active, relative, temporally-defined,* and, of course, *spread.*[21] It is also utterly physical and fits both with the physical world and with one's experience.

The nature of the object can be revisited in causal and temporal terms. Objects take place relatively to the casual circumstances our bodies contribute to. They would not exist, as they do, if our bodies were not there; they are not internal to our bodies. They are akin to the relative velocities of things surrounding us. Such relative velocities depend on our own speed and direction, yet they are both physical and relative to us.

What we label "an object" is a physical occurrence that repeats itself whenever we put our bodies in the proper circumstances. As a result, even though we believe the same object perdures, what perdures is not the object but a set of circumstances that are favorable to the occurrence of a series of identical objects. A key difference is that the model of the object I outline here and the traditional notion is that the spread object *takes place* while the traditional Galilean Object *exists.* The latter is supposed to wait there until someone or something interacts with it.

The spread object is like a fridge light: when I open the refrigerator door, the light switches on. Whether the light is on when I am not looking inside makes no difference to what I see. Indeed, I know that the light switches on when I open the door because I have some alternative knowledge as to the causal structure of the fridge. Because of my knowledge of the fridge, I assume the light is turned on only when the door is open. This does not entail that the fridge light is in any way a mental entity or that it supervenes on any fancy emergent property. It shows only that the switching of the light is a physical occurrence caused by the opening of the door. An earthquake might shake the fridge so much that the door will open without any human intervention. In everyday life, though, a human being is a favorable cause of the switching on, usually more frequently than earthquakes. Similar to the fridge light, the spread object takes place whenever a body with certain characteristics is available. The body we need is not metaphysically special. The body is just a physical aggregate that allows certain causal processes to reach an end. In all respects, *the body is just yet another* object. For instance, eyes are able to trap processes made up of light rays.

The human body offers the favorable causal circumstances to fulfill the causal role of those objects that, unsurprisingly, we meet during our life. Like the fridge light, the objects we are familiar with take place whenever a human body is causally coupled with certain physical stuff. Such stuff is not yet the object one perceives. The stuff becomes the object we perceive when it can act as an object thanks to the right favorable causal circumstances. The object we perceive is the physical cause affecting our body. It is not, however, the effect inside our body.

A big advantage of this account is that neither additional mental entities nor relations of any kind are necessary. Physical objects are everything one needs—i.e., the stuff the environment is made of and one's body. The object takes place when body and world have causal intercourse. Why should we add anything else? Our bodies are like the switch in the electric circuit. Whenever one's body is in a certain place, the stuff that makes up the world triggers a causal process, and thus the object, as the cause, takes place. The current—the causal process that brings the object into existence—flows. In this account, the human body and the brain do not concoct anything inside themselves. The current and the light are not concocted by the switch. Similarly, the object is not concocted by the body. Likewise, nothing is inside the switch and nothing is inside the living beings, regardless of Galileo's intuitions. The body only offers additional causal circumstances to its surroundings—just like every other physical entity. However, human bodies are key for human beings. Trivially, the world we are familiar with is the one that human beings bring into existence. My world is made of colors rather than of electromagnetic fields, because colors are the kind of things that produce an effect thanks to a human body and brain; electromagnetic fields do not. Such a world is not a hidden, invisible, private, mental, arbitrary, phenomenal, subjective domain. On the contrary, it is both physically observable and sharable. In fact, we all partake in the same world.

The object we perceive is not mind dependent. However, it is brain (and body) dependent. The kind of dependence is not constitution but causal contingence. Brain dependence does not imply constitution. The object is constituted by the external world and singled out by a causal process. The light is lit by electricity but it is only possible thanks to the switch. The picture is, admittedly, reminiscent of Berkeley's view with three key differences. First, the object is not created by a mental act—the object takes place because the proper physical conditions are met—the body being one of them. Second, no pre-existing subject is needed. Third, the object is a physical one. Since the body is just yet another object, the dependence on the body is nothing but object-dependence.

But, one might ask, aren't objects there when no human being is around? Consider this case. In Italy, one of the masterpieces of the Renaissance is the façade of Leon Battista Alberti's Basilica of Sant'Andrea in the beautiful town of Mantua. As an object, the façade needs a human body made of—in addition to other essential machinery—a visual system with eyes plus the proper neural structures. Because of such a body, the façade takes place as a complex, geometrical set of entities that are located where everyone takes them to be: namely, on the front of the main church in Mantua and not in the viewer's brain. However, in Mantua, Alberti's façade takes place whenever one looks at it, and thus all visitors go home satisfied. When pedestrians and visitors stroll across the square, they cannot help but see the façade. In fact, visitors cannot visit Mantua without their bodies. Yet such a reliable presence does not entail that the façade exists when there are no human bodies around. Like the fridge light, when there are no human bodies around, as on a cold winter night, no façade exists—there is only the stuff that, in causal conjunction with a human body, will make a façade. The stuff, which is there when no one is looking at the church, is, so to speak, God's view of the building.[22] It is something we cannot experience directly because the very existence of our body brings into existence a relative version of it—namely, the spread object (in this case, the façade). The spread object is not a private mental object. It is not metaphysically private. It is only as private as a hot dog is private—only one person can eat it. It is the façade that everyone sees when one takes a walk in the main square in Mantua.

In the physical world, "taking place" is never a lonely business. It always requires two physical entities or more. The ball can hit a wall only if both a ball and a wall are available. The tree can cast a shadow on the ground only if a tree, light rays, and a ground interact together. The key opens a lock only if the right lock exists. The propeller propels only if there is water. And so forth. The core idea is that what we perceive exists because a causal process made it actual. Such a process is usually called "perception." Perception is just a physical process. No mental aspect is required. However, such a process gained a special name because it allows the existence of the objects of which our experience is made. *Perception does not generate mental percepts but singles out external objects.* In other words, objects are not passive targets of perceptual systems. Neither are objects phenomenologically dark because there is no such thing as a phenomenal world. Rather, objects are actual causes whose outcome is some event inside our body, normally inside our nervous system. However—and this is key—the physical objects we experience are active causes thanks to our bodies. Our bodies are among the conditions that allow the environment to take place as the objects we experience.

We do not perceive the object with God's eye. As Galileo, who was a Platonist, would have appreciated, *we cannot perceive the object as though our body were not there. The object we experience—that is, our experience— would not exist unless our body was there too.* The object we perceive is the one whose existence is causally carved out relative to our body. In the physical continuum, countless phenomena take place without being part of our experience: infrared and UV light, cosmic particles, magnetic fields, radioactive particles, and so forth. In contrast, certain phenomena require human bodies to take place. Our body is physical and thus makes a physical difference for its surroundings. Our body acts as a switch, a lock, a dam. The world is different because of it. Such a difference con-stitutes our experience and our world—experience and the world being the same. *We experience the world that takes place because our body is at a certain location and at a certain time.*

The human body is not metaphysically special. It is yet another object akin to many other, less glorified, things. My running shoes, the wall, the half-empty mug, and the washed out printed-copy of Botticelli's *Primavera* hanging from the wall are all portions of the physical continuum. However, they have different causal enabling skills. For instance, my running shoes are causally oblivious to the chromatic patterns on the Botticelli print. In turn, molecules of bu-tyric acid are immaterial for my retinas.[23] Likewise, many objects exist regardless of human bodies, but not regardless of proper causal processes. Physical processes take place everywhere and bring into existence everything that exists—or, to put it more accurately, that takes place. *At any given time, the sum of all causal processes fixes the actual ontology of the world.*

Human bodies are not ontologically special. They are objects among objects. For us, though, human bodies are paramount, since they are the objects that bring into existence those objects—apples, cars, faces, musical tunes, constellations—that are one and the same with our ex-perience and thus with us.

As it happens, the human body is equipped with one of the most complex causal structures known—a working brain connected with the sensorial apparatus of a working human body. Because of such a complexity, a wide range of events and objects take place: patterns, shapes, relations between proximal reflectance distributions, patterns of air pressure waves, and so forth. A plethora of objects takes place because of human bodies. The complexity of our brain is necessary to bring into existence the complex objects we are familiar with.

What takes place because of one's body is not special. It is like the fridge light. It takes place because the door is open—there is no big metaphysical mystery. The visual object I call "red apple" takes place on the table because of my visual system and, right now, it would not take place if my body were not here. However, if, by sheer chance, the same chromatic patterns produced an identical effect on, say, my pen, the theory of spread mind would not falter in describing what takes place. The object "red apple"—which is part of my world right now— would exist thanks to the pen rather than thanks to my body. Consider this analogy. Due to contingent causes, a human being is usually re-quired to turn a key inside a lock. However, because of unusual cir-cumstances, it might happen that a key encounters a lock and opens it. Such an unlikely case does not entail deep metaphysical revelation. The turning of the key, which usually requires human intervention, might occur because of unusual combinations of causes.

.   .   .

First and foremost, objects are relative objects. They are relative to other objects rather than, as is the case with idealism, to subjects. This is key. The kind of relativity we are considering is a pure physical no-tion. It is just like the notion of relative velocity—something that does not require one to step outside of nature.

In the case of human beings, we experience the objects that exist relative to our bodies, which in turn are nothing but objects. Our bod-ies are objects at the center of huge causal networks that bring relative worlds into existence. Such worlds and our conscious minds are one and the same.

To get the gist of the idea, consider once again the notion of relative velocity and that of reference frame. The body is a complex reference frame. It plays just the same role of a reference frame for relative velocities, only it is much more complex. However, a body is nothing but a reference frame.

Consider this analogy. I am driving my Ford along the highway with a velocity of 50 mph. John is driving his Tesla at 70 mph. Francesca is driving her Ferrari at 80 mph. At the same time, a truck moves on the slow lane, lazily moving at 40 mph. Of course, all such velocities are relative to the ground. They do not exist autonomously. In everyday life, this fact is rarely mentioned but it is well known by every schoolboy. Let's focus on the truck. What is its relative velocity? It will be 10 mph relative to my car, 30 mph relative to John's car and 40 mph relative to Francesca's. Thus, for each driver, the velocity of the car is different, it keeps changing as we accelerate or decelerate, and it is also private in the sense that we cannot experience the truck's velocity.

Velocity is a simple physical property. It is easy to visualize the connection between the three cars and the truck. Each car corresponds to a reference frame relative to which the truck has a different relative velocity. Each velocity of the truck exists only relative to one of the cars. The notion of reference frame is only an abstraction to refer to the concrete existence of another object—e.g., a car.

In the case of other properties—such as weight, length, color, shape, texture—the connection between external objects and human bodies is much more articulated and more difficult to envisage. Yet, it is not different. Bodies plays the role of *complex causal reference frames*. Bodies bring complex relative properties into existence just like cars moving on the highway bring relative velocities into existence.

As the same truck has different velocities for each car on the highway, the same object has different colors, sizes, shapes relative to different bodies. The same board may be white relative to my visual system and pale red relative to the visual system of someone who has undergone chromatic adaptation. The same mug may be hot to my hands and lukewarm to Francesca's hands. The same truck can move at 40 mph relative to Francesca's car and 10 mph relative to mine. No need of any mental property, though. Everything is physical and relative. The notion of relative physical property encompasses all features of subjective experience. Subjectivity is no longer needed.

Thus we can bring back the old question: if a tree falls in a forest and no one is around to hear it, does it make a sound? Compare the question with this one: if a particle is alone in the universe and no other particle is there, does it have a velocity? The reply is negative. A lonely particle has no velocity at all. It is neither in motion nor still. The question becomes meaningless. Since Galileo, velocities are all relative velocities. My claim is that all physical properties—at least those that partake our experience—are relative properties. The world, which we experience and we live in, is relative, made of objects relative to our bodies. Since the properties of the objects that surround us have no absolute existence, they are different for each of us and are also fleeting and private.

•   •   •

Existence is relative. Since objects are bundles of relative properties, objects are relative objects. They are also actual insofar as relative existence always needs to be embodied by an ongoing causal process. Existence is relative and actual.

The notion of relative and actual existence was significantly raised at the very beginning of western philosophy by a mysterious character in *The Sophist*, one of Plato's dialogues. An unnamed philosopher, nicknamed "the stranger" highlights the connection between existence and causal efficacy:

> I am saying that it is whatever possesses a specific power either to naturally do anything whatsoever to another or to be affected in the smallest way by the slightest cause. Existing is nothing else but power.

In this seminal passage at the dawn of our civilization, power is nothing but the fact of existing relative to another object. The being is the causing and, in turn, a causal connection is the way in which something exists relative to something else. The world exerts its effects relative to our bodies. If our bodies were different, the effects would be different too. If the effects were different, the relative causes would be different too. The relative nature of existence is embodied by means of causal processes. Objects do not have absolute properties. They are carved out by their relation with other objects. Existence is not absolute but relative to the capability to produce effects—"nothing else but power." In turn, producing a certain effect requires the proper causal circumstances. The objects we perceive are always the causes of effects in our bodies and particularly in our brains. Bodies are like causal lenses that bring together multiple paths and thereby create causal unities, which we call objects.

The connection between existence and causation was re-formulated in 1920 by philosopher Samuel Alexander, in a passage in his work *Space, Time, and Deity*:

> [Epiphenomenalism] supposes something to exist in nature which has nothing to do, no purpose to serve, a species of noblesse which depends on the work of its inferiors, but is kept for show and might as well, and undoubtedly would in time be abolished.

The passage states that if something is epiphenomenal—i.e., it doesn't have any causal power—it does not exist. Exactly what the mysterious stranger claimed more than two millennia ago. More recently, in 1993, the influential philosopher Jaegwon Kim stated that "To be real is to have causal powers." Eventually, in 2005, Kim stressed the same point in even more radical words: "To deprive something of causal powers is to deprive something of its existence." Along the same line, in 2001, the philosopher Trenton Merricks expressed the view that "for macrophysical objects, to be is to have causal powers," a principle he named the Eleatic Principle—the attribute "Eleatic" is a tribute to the aforementioned Platonic stranger, who stated he came from the Greek town of Elea.

In short, both in science and in philosophy, many have wondered whether existence and active causation are one and the same. In addition, I want to emphasize that existence is relative to the causal circumstances offered. Existence is relative then. Do we ever experience anything that does not cause an effect? In practice, we do not and we cannot. In theory, I do not see how. The world we experience and we live in is made of actual causes. We do not experience idle dispositions or causally inert entities. We experience things that take place and cause effects. Everything we perceive is something that, when we perceive it, causes something. I dub the cause whose effects take place now as the *actual cause*. We experience an actual and causal world—i.e., a world made only of actual causes. An object is an actual cause since an actual cause is always relative to a physical system that allows its occurrence. An actual cause is also an actual relative object—i.e., a spread object.

·   ·   ·

The spread object is based on a causal account of objects. An object is the actual cause of an effect. Such an occurrence carves out a mouthful of nature every time one experiences it—experience is a causally carved out object. The word "experience" might be misleading since it suggests that something "mental" is the cause of the occurrence of something physical, namely, "the object." In this case, though, since experience is one and the same with the object that is causing an effect, the relation between objects and experience is no longer metaphysically demanding. *Phenomenal properties are physical properties.*

The spread object confines the traditional Galilean object—the one that is allegedly there when we do not look at it—to be only an ideal abstraction, useful but misleading. In fact, such an object is something that does not do anything. It cannot be experienced by anyone. It is beyond our causal grasp. It is invisible and phenomenologically dark. In short, the Galilean object is suspiciously akin to Platonic ideas disguised as modern entities. In fact, it is a useless hypothesis since no one benefits from it. It is like the fridge light when the door is shut—a useless hypothesis and, if switched on, a waste of energy and money. We gain no advantage from assuming the existence of objects that we perceive when we do not perceive them. Of course, this is no admission of idealism. Spread objects are not mental entities created by minds (aka persons, aka subjects). Spread objects are actual physical causes, taking place relatively to other physical objects such as our bodies. Crucially, our bodies are the only thing we cannot be. Thus, in a strong sense, the objects our experience is made of are not created by us, but rather by our bodies, which are not us.

To recap, the spread object is an actual physical cause that is identical with one's experience of the object. I do not have an experience *of* or an experience *about*. Rather, I have an experience that is *identical with* an object. I am the object I experience, or rather the set of objects I experience. The object is the thing I am when I experience it. I don't have literally an experience *of* red, *I am a red object*.

In causal terms, the most fleeting experience is not different from, say, Mount Elbert. They are both causes. The only difference is that usually the conditions that allow the mountain to take place last much longer than those that allow the existence of, say, a shade of red in a rainbow. These differences do not change the physical nature of the phenomenon, only the lifespan of the conditions that allow it to happen. However, both mountains and shades of red take place anytime someone looks at them. Such a causal account does not contradict any physical law, any empirical data, or any phenomenological report. The stuff the Rocky Mountains are made of is there all the time, but the object I call Mount Elbert, which I see when I go there, takes place only when my body is there.

Are such objects types or tokens? In this account, everything is a token. Consider this case. Today, at 11 a.m., I go from Kendall Square to Harvard Square by train. Which train do I take? The Red Line. Then I walk back. I have lunch and, at 4 p.m., I move back from Kendall Square to Harvard Square. Which train do I take this time? The Red Line. On the one hand, I take the same train. On the other hand, I do not. In some sense, the Kendall Square to Harvard Square Red Line train exists. However, it is also true that today's 11 a.m. train is different from yesterday's 4 p.m. train. The key point is that, between the alleged Galilean object and the real object, a difference analogous to that between the ideal Red Line and the 11 a.m. train from Kendall Square to Harvard Square holds. The former is abstract, atemporal, and absolute. It does not carry any passengers. The latter takes place every time I board it. It is temporally located. It consumes energy. It takes time to complete, due to nomological constraints. It is dirty and full of passengers. The exact times of its arrival and departure can vary. The former train is only an immaterial abstraction concocted by linguistic practice. The example of the train is informative. I can say that I take the same train or that I take two different trains. I can say, "I take the same train every morning," and that is fine. Alternatively, I can say, "I took the 11 a.m. train on June 24, 2014 and it was much cleaner than the 4 p.m. train on June 25," and that is fine too. The former case is akin to the Galilean object and the latter to the actual one.

The spread object is like the train one gets aboard between two stations at a specific time. If one did not pay for the ticket, in some sense, no actual train would occur: only gears, wheels, engines, and carriages. However, if one pays for the ticket, in this metaphor, the train will occur. The object is singled out and a process runs from the stuff in the environment up to your neural structure. The causal run from the environment up to your nervous system singles out a certain object-train in the world. Every time one looks at a certain causal configuration, the corresponding object takes place. A train starts and gets to its destination. A circuit closes and a fridge light is switched on.

The traditional notion of the Galilean object is popular because of its undeniable practical efficiency. Likewise, when I watch a movie, it makes sense to refer to a character, even though the character we love is just a sequence of frames, each one physically separate from the others. The causal intercourse between the stuff in the environment and human bodies pulls into existence the familiar objects that are our experience.

Everything is carved out in causal terms. Since the body has no special role, many causal processes—and thus many objects—take place also in the absence of human bodies. The interplay between processes in which our body has a direct role and processes in which our bodies have no direct role has side effects as to how we categorize objects. Usually, if an object does not require anything as complex as a human body to take place, it is taken to be more "substantial." However, it is a matter of degrees rather than an ontological gap. The more easily an object takes place, the more it is taken to be real and hence not subjective. In practice, though, human skills are crucial to bring into existence many familiar objects. Human bodies carve out the physical continuum into different—partially overlapping—subsets. Imagine that Emily is still sitting next to me in the Boston library. There are objects that take place thanks to Emily's body, objects that take place thanks to my own body, and objects that take place thanks to other physical systems—the floor, the table, the lamps. These sets of objects partially overlap. What I see overlaps with what Emily sees. No matter what one's perceptual system singles out, additional processes are always taking place outside. They bring further causes into existence.

Consider three crosses endowed with apparently different degrees of ontological autonomy (Figure 4).[24] The gray cross on the left seems very "real." The other two (center and right) seem to be arbitrarily created by the beholder (a cross of "n" and a cross of prime numbers). Yet such a difference is parochial. If we possessed hard-wired receptors for prime numbers (something easily achievable using a Character Recognition Module and a mathematical rule), the last cross would appear as conspicuous as the first one. The alleged different degrees of autonomy depend on the effort the brain must make to single each cross out. Yet, the three crosses take place because the stuff they are made of allows a certain object to take place, if coupled with a brain. A cause requires a proper physical system to take place.

Our experience of the world is made of actual entities. The physical world we live in is made of actual entities too. I experience something that takes place. I do not experience what the world might be—I experience what the world is. I am the actual occurrence of a world.

u	u	n	u	u
u	u	n	u	u
n	n	n	n	n
u	u	n	u	u
u	u	n	u	u

30	8	17	25	28
18	15	11	9	14
31	3	5	23	29
10	22	19	4	20
16	8	7	6	12

Figure 4. Three objects. Are they real to the same extent?

. . .

And yet one may wonder about the inconstancy of our experience. Is it not an assured fact that the way in which the world appears to us keeps changing? Isn't the world as we perceive it subject to change due to the subjectivity of our unreliable perception? Do we not perceive differently because of our different senses? Do objects not manifest themselves differently based on the sensory modality? Is the same object not different when perceived through the modalities of touch, smell, taste, hearing, and vision? Finally, are the tactile and the visual apple not different modes of presentation of the same apple? Is this the usual mistrust Galileo and science have cast over our senses? As a matter of fact, all physical properties are relative and are thus subject to the same kind of continuous variations.

The reply dictated by the theory of spread mind is unambiguous— *we do not perceive differently the same object, we perceive different relative objects*. Rather than perceiving an object in a certain way, a sensory modality singles out a group of objects. As stressed, the spread object is causally defined. Consequently, different sensory modalities—each corresponding to a different causal process—single out different objects. Different sensory modalities are different clusters of many causal processes. For instance, vision lumps together causal processes that exploit light rays as their medium. Hearing lumps together causal processes that exploit waves of air pressure. And so forth. In the presented account, a sensory modality is a set of homogenous external causes rather than a set of homogenous internal phenomenal characters. *Seeing an object differently is tantamount to seeing a different object.* Each sensory modality is a set of objects lumped together by the fact that they are all relative to the same sensory organ. Once again, phenomenal experience is a useless hypothesis once the relative existence of external physical phenomena is taken into account.

Since the objects we perceive exist relatively to the causal circumstances offered by our bodies, changes in our bodies or considering different parts of them are going to bring into existence different external objects out of the same physical continuum. If our bodies change, the world that they bring into existence—the objects that exist relatively to it—will change too. Objects are relative.

When my body is in front of the stuff that is the cause of a certain causal process, a certain combination is singled out of the many possible ones. For instance, what I call a "visual red apple" is singled out relatively to my visual organs. The "visual" here does not refer to any internal mental character, but to the fact that only optical phenomena are involved. Visual organs single out optical causes—namely, visual phenomena. Of course, many other processes can and do originate from the same stuff. When my hands touch the stuff on the table, a tactile apple is singled out (no more red here). When my tongue licks the apple or my teeth bite it, other causal processes single out different objects. As mentioned above, the stuff the apples are made of is real and constrains what my body can bring into existence. No matter how many times I look at the visual apple or how many times I touch the tactile apple, I can only eat one apple. All these apples are different relative actual objects singled out by different causal circumstances offered by different parts of my body. In fact, the causal process, which is my digestion, is destructive. It is only a contingent fact that most processes involved in perception are not destructive.

Eating is a good pragmatic criterion for existence. Nonetheless, every time I look at the stuff on the table, a new apple is the cause of the light rays that enter my retina. Every time I touch the apple, a new apple is the cause of the pressure on my fingers.[25] And so forth. Senses are causal gates for different objects, whereas such objects are only a few of all the objects that might be singled out of a certain environment. The bottom line is that sensory modalities neither add nor own specific qualities—*a sensory modality is a group of objects.* Is such a multitude of objects a problem? Not at all. Consider the case of relative vs. absolute velocities. Of course, the notion of relative velocity implies that each object rather than one absolute velocity has umpteenth relative velocities. Is this a problem for physics? Of course not, because each relative velocity is absolute relatively to the pair of bodies under scrutiny. The relative is absolute in its own scope.

A similar account can be outlined for modes of presentation such as when I see the same apple either with my glasses on or off. With my glasses on, the features of the apple appear sharp. Otherwise, the apple's features are blurred. Do modes of presentation compel us to admit that how an object appears is different from what an object is? Of course not. The blurred apple is no less physical than the sharp one. Both apples are physical. Tantamount to how different sensory modalities are objects singled out by causal processes, different modes of presentation are further objects singled out by further causal processes. The blurred and the sharp apple are two distinct physical objects overlapping only partially.

When I take off my glasses, I see the blurred apple. *Do I see the same object in two different ways, or do I see two different objects that I have practical reasons to address as though they were separate versions of the same one?* I defend the latter option. One exists relatively to my eyes with the glasses on. The other exists relatively to my naked eyes. Differences in experience correspond to different external objects. In this case, different objects exist relatively to different circumstances instantiated by my body in different configurations.

As it happens, the blurred and the focused apple are two separate overlapping objects. When I perceive the blurred apple, clusters of points on the apple's surface are lumped together. Such clustering is what one describes as blurring. On the other hand, when I perceive the focused apple, things are not so different. Each point of the apple is still the sum of many subareas that my visual system lumps together. If I had a better retina, I could see another, even sharper apple.

Another example is a checkerboard of red and blue squares. From a sufficient distance, it is a magenta board because a human retina cannot distinguish the squares. In a very strong physical sense, the board is both a magenta board and a red-blue checkered board. They are two relative objects, each exists relatively to the proper physical system, which in this case is a human body at different distances. Neither version of the board is more real than the other one. Nor are they located where the beholder's body is. The beholder offers the causal circumstances and its position and visual skills are a causal condition for the existence of either a magenta board or a red-blue checkered one.

Changing the beholder's body changes the causal circumstances, and thus changes the relative spread object that is brought into existence. If there were two human beings at different distances, there would be two objects, two boards with different colors. Is this absurd? Why should it be? If I have two cars approaching a lamppost at different speeds, doesn't the lamppost have two different relative velocities? Yes, of course. Existence is relative.

Spread objects are, so to speak, cheap. Whenever I experience differently, I experience a different object. Likewise, when I take my glasses off and see the blurred red apple, I do not need to assume I see the red apple by means of a mode of presentation that blurs the fruit. My body has changed and, consequently, the relative object my body brings into existence has changed too. First and foremost, the perfectly focused red apple is an abstraction, too. What we call a perfectly focused red apple would be a hopelessly blurry red apple if we had, say, a retina of tenfold acuity. By and large, the notion of a focused red apple is parasitical to the way in which our visual system works. Normal vision singles out a familiar yet arbitrary choice of objects. Blurred vision, astigmatic vision, and super vision are possible choices. None of them is ontologically better. None of them is closer to the ideal object. They are all real physical objects and all equally arbitrary and relative to one's body. Monet's landscapes are as much realistic as Dürer's still lifes. Blurred apples are as physical as focused apples. Blurred and focused apples are only two objects amidst countless possible ones. Each object is singled out by a specific causal process.[26]

The main reason why some authors prefer to consider the blurred apple and the focused apple as though they were two modes of presentation of the same object rather than two objects is the same one mentioned above. There is just one edible apple—between stomach and eyes, hunger has more authority than sight.

Strange as it seems, modes of presentation are not different ways for an object to appear. They are multiple objects which we lump together for practical purposes—eating, buying, owning, and trading. When practical purposes admit different behaviors, our ontology adapts. For instance, I pay for visual objects that happen every time I go to a movie theater.

The same approach holds for viewing something from different angles, or using one's sensory apparatus in different ways. I touch an apple either with my fingers or with the back of my hands. I look at it from above, from the side, from far away, or from a few centimeters. Every time I have a different experience, I single out a different object. If the body changes, the relative object will change too as is the case with relative velocity.

Sometimes, lumping many objects in manageable clusters is less obvious. Imagine that every morning when I wake up, I see a cat at my door. I call her Lucky. However, unbeknownst to me, my mischievous neighbor brings a different specimen onto my doorsteps every morning. Therefore, in reality, after a year, I saw 365 different cats. Nonetheless, in my experience there is just one cat. Lucky is a composite cat that I make up out of 365 different cats that my mischievous neighbor brought to my doorstep. This case might sound bizarre but it may happen nonetheless. It is akin to what we do by composing Han Solo from thousands of film frames. In my world and in my experience, nothing distinguishes an amalgam of 365 different cats from a cat. Likewise, whenever I see an apple from a different perspective, my body singles out a different apple. However, I believe I have always seen the same apple, mostly because, in the end, I can eat it only once.

·   ·   ·

The causal nature of objects entails their temporal nature too. In fact, in our universe, causation requires physical processes and, in turn, physical processes require time. Thus the causal nature of objects and wholes entails that they are spread in time. Objects are then intrinsically temporal, which is consistent with the nature of experience. Both consciousness and objects take place. They partake in the becoming of the physical world. Another gap between experience and the world can be closed. Since causal processes spread in space and time, an object takes place in time too. Of course, an object is not smeared over the time span of causal processes, nor is the object akin to four-dimensional worms. The object remains spatiotemporally located where its parts are. If the object that I perceive is, say, the star Sirius, the object will be temporally located 8.6 years before the corresponding activity in human brains.

The suggested causal account can be applied to all cases I have been able to think up—such as constellations, faces, words, tunes, chairs, stars, planets, books, living beings, persons, and neural firings. Such objects bring together elements that are spread both in space and in time. Consider a constellation. The constellation is where and when the stars are—or where they were. Since the stars are at different distances from the earth, the constellation is a physical object spread over a huge spatiotemporal span.

A causal account of objects allows us to merge experience and reality seamlessly. Objects are no longer static entities. They exist thanks to relative causal processes, but they are not such processes. The problem of objects shifts from a timeless perspective to a temporally oriented view. An object is the cause of a causal process. Consequently, if time froze, no object would exist: no time, no objects. This has interesting consequences.

Suppose time halts. According to a common opinion, if time stopped, everything would continue to exist, albeit frozen as in a snapshot. Raindrops and snowflakes would stop motionless in mid-air. Cars and bystanders would stay motionless. Patterns and shapes would remain where they were. The universe would freeze, but it would be ontologically unaltered. Such a popular narrative is both wrong and misleading. *A temporally frozen universe would be empty.* If time froze, everything would disappear. In such a timeless instant, no sound would occur since sound is made of sequences of air pressure waves. Likewise, utterances would disappear too since sounds are made of spoken words. Neural activity would cease to exist since it is made of sequences of chemical reactions distributed across time. Even light would vanish since light rays travel in time. The world would be pitch dark. Finally, objects would not exist since they require time to take place. There would be no patterns, no light, no sounds, no shapes, and no objects. And, not surprisingly, no conscious experience. A frozen universe would be changeless, empty of experience and objectless. If time halted, not only would the universe be dark and silent, it would also be empty.

The world is made of objects taking place at different temporal lengths. At the same time, different objects are spread across different time spans. An interesting example is offered by the human perceptual system in which different properties takes place in different instants.[27] For example, movement takes more time to produce an effect than color. A bright spot produces a very fast response while a face takes a relatively longer time. The world we live in is not an instantaneous point in time.

Perduring objects needs to repeat continuously because they do not exist but instead take place. They ought to repeat to produce to exist again and again. Whenever objects endure, it is because of the repetition of a series of seemingly identical objects. Consider again the analogy with film frames. To show a movie of a still object, a series of identical frames has to be projected. If, so to speak, we need the same object again and again, we should bring about the same causal process. To do so, we would need the same context. This is just what happens whenever we look at a static pattern or a still object. Whenever we look at it, a new causal process takes place. This is, once again, the fridge light model. The still object we see is a series of identical-yet-numerically-different film frames. A perduring object is akin to a movie about a still object.

The temporal structure of objects—and thus of nature—is key to allow a seamless identity between objects and experiences to occur. Both objects and our experiences—the two being the same—consume their existence in a moment of becoming.

· · ·

If effects are among the conditions that bring an object into existence, since they take place after the object, do they change the past? This apparent causal and temporal loop reveals the unsaturated and potential nature of reality until future effects seal a portion of the past. It is a kind of causal and temporal entanglement that resonates with what we are discovering by means of quantum mechanics. The present fixes the past.

The catch is that the object does not exist until it will produce effects. It is as though the ontological status of the object remains open until its effect will occur, no matter how long it takes. Any object is a possibility that, if it does not get to an end, will remain just a possibility. Possibilities do not grow old. This notion of suspended existence echoes nicely T. S. Eliot's lines in *Burnt Norton* (1936):

> *What might have been is an abstraction*
> *Remaining a perpetual possibility*
> *Only in a world of speculation.*
> *What might have been and what has been*
> *Point to one end, which is always present.*

However, when the effect takes place, the object will come into existence at the time where its parts where. Thus, for the object, no time will elapse until the effect occurs since no object exists yet. What happens in the present redefines the past. It does not make sense to speak of a time internal to a causal process.

Importantly, the theory of spread mind does not require any backward causation. Nothing goes backward from the effect to the cause. The object will remain indefinite until its effects occur. The effect changes the past, by bringing a cause—the object—into existence.

Consider another analogy, the winning lottery ticket. Suppose John buys a lottery ticket on June 14, 2015. The drawing will take place on the night of December 31, 2015, at midnight. Until that day, John has not bought the winning lottery ticket. He has only bought a ticket. Luckily, on the night of December 31, John's ticket is picked. Until that moment, unless one assumes a very deterministic version of the universe, all odds were against John buying the winning ticket. Nonetheless, his ticket won. As of January 1, 2016, John bought the winning lottery ticket on June 14, 2015. Nobody can deny that John bought the winning lottery ticket, six months before, yet it only becomes the winning lottery ticket after it won. It is as though the tenseless proposition "John bought the winning lottery ticket on June 14, 2015" was open until December 31 and true afterwards. In other words, future events complete and fix the ontology of what has happened before. Does it seem paradoxical? Only because we are used to conceiving of causes and effects as though they were aloof monads encased in watertight instants in time.

An object does not exist autonomously, but only relative to another object that is necessarily located at a subsequent moment in time. Existence is relative.

Another remarkable example is offered by rainbows, which are beautiful and revealing objects.[28] Consider a simplified rainbow, which is made of droplets floating in the air. A stream of parallel sun rays collides with them. As a result, each droplet reflects a divergent stream of colored rays. Depending on their location, different observers will select different combinations of droplets, since they are in different positions in space and thus they will single out different rainbows. If no observer were there, the reflected rays would not cause any joint effect and would continue to travel, getting lost somewhere. They would miss their only chance to act together. Thus, no rainbow as a whole would exist. Yet the rainbow is there in the cloud. It is neither in our head nor in our mind. As the poet Gerard Manley Hopkins observed in his poem *It Was a Hard Thing to Undo This Knot* (1864),

> *The rainbow shines, but only in the thought*
> *Of him that looks. Yet not in that alone,*
> *For who makes rainbows by invention?*

The rainbow is not in the thought alone. The rainbow is in the cloud. It is a physical object that can be photographed. Yet, its existence is relative to another object—be it a human body or a camera. In the cloud, an almost infinite number of possible rainbows might take place. Yet, only a very limited number of them succeed. Only those that interact with the proper physical systems have a chance to exist. The actual rainbows are the ontological winning lottery tickets and the observers' bodies are the drawing processes. The cause depends on the effect to be a cause while, obviously, the effect depends on the cause.

· · ·

The spread object has its own quirks—namely, the relative, casual, and temporal location and duration, the recurring nature, and the causal dependence on the body of the observer. Crucially, the *spread object does not conflict in any way with empirical evidence.* Whether the object I experience exists when I do not experience it—like the fridge light when the door is shut—is immaterial. *We cannot experience what the world would be if our bodies were not there.* Likewise, *what we experience is what the world is when our body is there.* It is of course possible to draw indirect inferences about the stuff that is not part of our experience— science does that and does it well. However, such stuff is not the object we experience. The authority of science is subordinate to that of experience. Luckily, science does not conflict with experience once one adopts the theory of spread mind. Experience is the cause of a process that builds what is physically there when one perceives it. *Opening one's eyes changes the surrounding world because it allows certain objects to produce effects and thus to exist.* Touching will allow other objects. And so forth. Perception is experience, which is, in turn, causal existence. Perception entails a change. Experience is the object causing that change. The spread object is *physically* carved out from the physical continuum.

The spread object is compatible both with one's experience and with scientific findings. Conversely, the Galilean object runs afoul of one's experience and requires that one does violence to the senses. The spread object is so close to one's experience that identity comes as a natural option. The spread object does not require any violence to one's senses. Additional entities—such as phenomenal experience, phenomenal character, modes of presentations, and sensory modalities—are no longer required. This is the end of the gap between the mental and the physical.

In the spread mind account, the picture is different. Objects and their associated experiences are no longer two separate domains. Of course, an object does not necessarily belong to one's experience. Each experience is an object, while not all objects are part of a mind. Minds are sets of objects we regard with particular affection because they are we and we are they. Under the theory of spread mind, the mismatch between experience and objects does not occur, since appearance and reality are one and the same. The identity between experience and reality must always hold. Such an identity is the key prediction of the theory.

There are two classes of objects: relative objects that are actual and part of one's mind and relative objects that are not part of one's experience. The second set is usually inferred by the first, through science and deduction. We infer the existences of the entities that are real but that are not part of anyone's experience—such as infrared light—by means of effects on the entity that we perceive.

The spread object shifts the burden of supporting the dynamic features of experience from the mind to the physical world. Objects, which are actual causes of physical processes, are all we need. The notion of the spread object overlaps that of experience. Both experience and objects are spread in space and time. Their boundaries are defined in causal terms. They overlap. Experience and objects are one and the same.

Experience and objects co-occur. They are just different ways to refer to certain occurrences inside nature. The object I perceive is not different from the object that is there. If someone eats the apple, the apple is gone from my experience too. The apple that sits on the table is the apple that is part of my mind. Why should I believe that what I see is somehow distinguished from the world? No special phenomenal coating covers the physical world. No mental paint or mental latex is needed.

There is no veil over reality. Reality presents itself as it is, *pace* Galileo. I will not be shy here: the object in my mind is identical with the object on the table. *Phenomenal properties are the physical properties of the spread object.*

The introduction of phenomenal properties has been akin to the invention of a mysterious quality that objects are expected to have when owned by someone. Imagine seeking a mysterious *ownness* that is instantiated by objects with an owner. Of course, such a mysterious property would be ineffable and invisible. It would also be empty notions.

The traditional contrast between phenomenally-empty yet causally-active properties and causally-inert yet phenomenally-rich properties is untenable. The hidden nature of the object is a fact we never experience. The Galilean object is a useless hypothesis that does not and cannot make any difference. Thus, we can finally set it aside. In 1904, William James famously stated, "Experience [...] has no such inner duplicity; and the separation of it into consciousness and content comes, not by way of subtraction, but by way of addition." Such an addition of a phenomenal layer is empirically ungrounded and conceptually doomed, not to mention carelessly prodigal from an ontological perspective.

Another cue that backs up the identity between experience and the world is that the properties of the world we experience match perfectly with those aspects of the environment that have causal intercourse with our body. The list of the physical properties we experience and the list of the properties that, thanks to the causal structure of our bodies, take place match perfectly. The objects, as they present themselves to us, are made exactly of those properties that can produce effects thanks to the structure of our bodies. We live in the subset of the world that exists because it is causally connected with our bodies. Such a subset is akin to the notion of *Umwelt* suggested by the ethologist Jacob von Uexküll: each animal lives in a subset of nature which is causally defined by the causal properties of its body. In his main work in 1909, later translated in 1957 as *A Stroll Through the Worlds of Animals and Men*, he outlined the notion of *Umwelt*, which is dynamic and actual and varies from moment to moment:

> We must first blow, in fancy, a soap bubble around each creature to represent its own world, filled with the perceptions, which it alone knows. A new world comes into being. [...] As the number of an animal's performances grows, the number of objects that populate its Umwelt increases. It grows within the individual life span of every animal that is able to gather experiences.

Similarly, I suggest that one's experience is identical with the particular subset of the physical world that takes place because of the causal intercourse with the body.

. . .

Can the chasm between phenomenal and physical properties be swept away so easily? I think so, since such a divide is grounded on questions begging assumptions—such as the conflict between appearance and reality—and on misinterpreted, empirically-shaky evidence—such as dreams, illusions, hallucinations, and misperception. I will address the latter issue in the following. Here, I indulge some more in the comfort of Shangri-La where each experience has—and thus is—an object. Consistently, the twin notions of the spread mind and spread object render the traditional separation between mental and physical unnecessary.

The notion that phenomenal properties are not part of the physical world is bizarre, to say the least. We have become accustomed to it because of philosophical practice, yet—one must admit—such a notion has not been so fruitful. Alleged phenomenal properties cannot be observed directly. Worse still, they cannot be compared against physical ones. I can check two lengths against each other. I can contrast two shades of red in my mind. If phenomenal and physical properties were metaphysically aloof, how could I match the red in my mind against a physical property in the world? Of course, I could not.

Has anyone ever been able to put a mental property next to a physical one and judge whether the mental property looked different from the physical one? Given how phenomenal properties have been usually defined, such a direct comparison is both empirically unsound and theoretically impossible. Phenomenal and physical properties have been split conceptually in such a way that they cannot be mutually compared. When one chooses which car to buy, one does not worry about the difference between the color of the car and the phenomenal color one sees. And, with good reason: such a quandary is the offshoot of analytical definitions rather than of natural obstacles. In everyday life, the world I live in is made of those very properties I experience. The distinction between a causally inert appearance and a phenomenally inert object is conceptually amusing but of no practical value. Every time I experience the world, the world is made of the same qualities I experience. Are the alleged mental qualities not the properties of the world I live in? My experience is made of people, houses, buildings, trees, cars, animals, rivers, mountains, and stars. My world is made of the same things.

Even more curious than the notion of phenomenal experience is the cognate notion that the world is devoid of the very qualities I experience. Only someone lost in the iron grip of a theoretical framework might accept such a notion. How could I possibly know that the world is devoid of qualities if I did not experience the world in itself? Who has ever perceived the world in such a way? Who could put forward any claim about what the world is when we have no evidence of it? How can anyone claim any knowledge about the nature of the world when the world is assumed not to be part of anyone's experience?

In contrast with such a traditional view, whenever I perceive an object, I experience that object. Even if my experience were not identical with an external object—and, according to the theory of spread mind, this never happens—my experience should be identical to something. Thus, in the world, something physical must have the properties of my experience. Consequently, the very fact I experience something, whether I am a spread mind enthusiast or an orthodox internalist, shows that the physical world has quality. I can be mistaken about the thing that has the quality of my experience—whether it be an apple or a neural process—but either way *something physical must be my experience and must have the properties my experience has.* I cannot have an experience of the world without having an experience! Furthermore, I cannot experience the world without my experience being something physical. If I am a physicalist, an experience must be a physical occurrence. The strange idea that the physical world is devoid of qualities is yet another side effect of the Galilean notion of the object. The quality-less object is not an object one can ever run into. It is a wireframe, an abstraction akin to the notion of center of mass. If one dug a hole and reached the center of the earth, one would not find anything. Consider again the Red Line train. No one can board it. Abstract train lines, centers of mass, and Galilean objects are not real objects.

Yet why are so many scholars confident that the physical world does not have the resources to be like the world we experience? Why do so many authors believe that the physical world is different from the mental world? Why do so many philosophers and scientists take for granted the separation between mental and physical properties?

I suspect that the main reason might be embarrassingly naïve. In the past, scholars and regular people alike often assumed that—whatever it is—the mind depends exclusively on the body. As a result, they looked for mental entities *inside* the nervous system. Since they did not find anything of the sort—and still do not—a widespread consensus emerged that mental properties are conveniently special. They must be invisible, private, unobservable, and causally inert. Thus, the separation is no longer a falsifiable empirical hypothesis. It is an unquestioned dogma. Once you accept the separation between phenomenal and physical properties, empirical evidence cannot challenge it. The very notion of a separation between the phenomenal and the physical is empirically unredeemable.

To be clear, I do not claim that experience does not exist, only that the traditional notion of phenomenal appearance—and qualia and their cognates—is an empty one. On the contrary, *experience exists and it is the world we live in.*

Since I have indulged in a minimum of historical wanderings, another connection worth mentioning is the tendency to place man at the center of the universe, if not astronomically, at least conceptually. I refer informally to any conceptual framework that gives a special position to human beings as a "Ptolemaic stance." It is all too easy to accept the cautionary tale of the demise of geocentrism and then fall prey to more subtle forms of geocentrism. The notion that the mind is somewhat special with respect to the rest of the physical world is a fantastic example of a Ptolemaic stance. Traditional substance dualism is another obvious example because it assumes that the thinking substance is special as regards the extended substance, which is the rest of the universe. The Cartesian mind is, in a metaphysical sense, at the center of the world of experience and knowledge, while the physical world loiters around dumbly and passively. Similarly, Blaise Pascal's thinking reed—however weak—in virtue of being the only thinking entity in the universe has a superior status. With rare exceptions, the mind has kept a central place in the phenomenal and epistemic heavens, flattering scholars. According to Franz Brentano, intentionality is a sort of hallmark that sets the mind apart from nature. And so forth. The stigmata of the mental are still present in the current formulation of the "Hard Problem," which promotes consciousness as a sort of ontological Robin Hood—something that science will never be able to catch fully. The mental or the phenomenal is often presented as something special that the physical world cannot cope with. According to the early David Chalmers, once all other problems have been solved, the question of the nature of consciousness will still be around—this is the so-called hard problem. According to Colin McGinn, consciousness will always evade a solution. The mind is the last stand against positivism. In the wake of such an aristocratic and Ptolemaic picture of the mind, the brain has been progressively puffed up with higher and higher praises and expectations.

In the meantime, the belief that neurons can do something metaphysically special such as propel the emergence of phenomenal properties has been reinforced. This is not to say that the brain is not an impressive biological machine. On the contrary, it is remarkable. Yet, believing that the brain can produce phenomenal properties is tantamount to believing that the brain can transform lead into gold—it is an incongruous hypothesis. If phenomenal properties were not physical, how could the brain foster them? No matter how much we are flattered by the idea that our brain has a special place in the universe, the brain cannot accomplish the impossible. In the same fashion, recent forms of reductionism of the mental appeal to some slight difference between neural processes and mere physical processes: computation, functional states, emergent properties, information, integrated information, cognition, enaction, and so forth. Such a quest is akin to looking for the philosopher's stone. A special ingredient is sought, be it a functional state, enaction, information integration, computation, or cognitive broadcasting. Of course, almost none of these authors are so naïve as to make the fatal mistake of subscribing to Cartesian dualism of substances. Yet, scholars and non-scholars alike enjoy the notion that the mind is a somewhat distinguished state of matter. In a physical world, the mind is special and brains share some glow, too.

The bottom line is that the conflict between the phenomenal and the physical is the last Ptolemaic stance in the history of science. In contrast, more democratically, the theory of spread mind flattens all differences between mind and nature, between phenomenal and physical, between experience and world.

·   ·   ·

The spread mind theory is neither textbook idealism nor naïve panpsychism. Both these views, at least in their traditional form, are different from what I flesh out here. The spread mind theory does not suggest that objects one perceives are mind-dependent. On the contrary, they are brain-dependent physical objects not located inside the brain. By contrast, according to the textbook version of Berkeley's idealism, everything is mind-independent. Everything exists only because the subject posits it as an idea to perceive—*esse est percipi ... aut percipere.*[29] Three differences are worth stressing.

First, while the popularized version of Berkeley's idealism accepts the separation between the mental and the physical, the theory of spread mind is a strictly physicalist view. In other words, everything is physical and everything has a causal role. Everything is an object.

Secondly, spread mind does not posit the subject as a necessary precondition for the occurrence of objects. Objects do not depend on the mind. Rather, the mind is identical with objects. The objects one experiences are brain dependent; a causal contingent dependence. Spread mind outlines a bundle theory of the mind akin to Hume's, except that physical objects play the role of impressions.

Finally, every object requires another physical system to take place. An object cannot take place in isolation. A shadow requires a surface. A key requires a lock. A face requires a human brain to observe it. Contingently, certain objects require physical structures that are embodied in human bodies. However, such objects are not mental entities. They are objects that take place thanks to other physical systems that happen to be human bodies. Everything is physical. By the same token, cars, trumpets, and cell phones require human beings to be manufactured. Cars, trumpets, and cell phones are not mental, however. Brain dependency is not mind dependency.

In many respects, the theory of spread mind is at the opposite philosophical end of traditional Berkeleyan idealism. The view can be summed up by the expression *perceiving is being (percipere est esse)*—to perceive an object is for that object to exist. If compared with the traditional views, the theory of spread mind is the only theory whose ontology is purely physical. The feeling is the being.

Panpsychism is committed to the assumption that phenomenal experience is something additional to the physical world. Such a hypothesis runs afoul of the spread mind theory, for which physical objects are sufficient. The popular notion of panpsychism is indebted to the dualistic notion of psyche by means of fictitious properties that, rather than being confined to one's mind, are prodigally given to all physical entities.

Physical objects carve themselves out from the physical continuum through interaction with human bodies, which are a part of the physical world. The apple on the table and the apple in my mind are the same apple. Having an experience of an apple is having that apple as a part of one's experience. The present account is thus a form of *no-psychism* rather than panpsychism.

3.

The Causal Geometry of Experience

> Sitting on the bus going up Shaftesbury Avenue,
> she felt herself everywhere; not "here, here,
> here"; and she tapped the back of the seat;
> but everywhere. She waved her hand, going up
> Shaftesbury Avenue. She was all that. So that
> to know her, or any one, one must seek out the
> people who completed them; even the places.
>
> **—VIRGINIA WOOLF, 1923**

F EXPERIENCE IS a matter of being one and the same with the physical world, how will this theory account for all cases of experience? In fact, the notion of a mental domain is convenient, albeit mysterious. Believing that what we perceive is just a mental image encourages lazy scientific habits since it places our experience outside nature. As a result, many scholars have found it convenient to explain perception in terms of hallucinations rather than the other way around. The prevailing model of perception can so be dubbed a *hallucinatory* model of perception, because it explains everyday standard perception as a case of reliable hallucination. It is a curious explanatory strategy because it explicates the normal by means of the unusual. Perception is seen as a form of world-matching hallucination.[30] A popular notion is that the world that surrounds us night and day is a sort of 3D movie unfathomably produced inside the brain. For instance, at a popular TED talk in 2014, the philosopher David Chalmers stated that

Right now you have a movie playing inside your head. It's an amazing multi-track movie. It has 3D vision and surround sound for what you're seeing and hearing right now, but that's just the start of it. Your movie has smell and taste and touch. It has a sense of your body, pain, hunger, orgasms. It has emotions, anger and happiness. . . . At the heart of this movie is you experiencing all this directly. This movie is your stream of consciousness, the subject of experience of the mind and the world. [31]

One cannot be more Cartesian than that. The world is conceived as devoid of qualities and conscious experience is concocted inside our mind, be it a brain or a soul. Unfortunately, nobody has a clue as to what such a movie is, where and what the theater is and, of course, who watches the movie! *Pace* Chalmers, nobody has ever found anything remotely resembling our experience inside our head. As the philosopher Tim Crane observed in 2017, "Looking at images of neurons under a microscope doesn't make the mystery disappear, any more than looking at a brain on the chopping board does."

Yet, the Cartesian model is successful because it is based on the separation between appearance and reality. The world is a sort of lifelong hallucinatory dream kept synchronous with everyday real life. Such an inverted explanatory direction is a clear offshoot of Descartes's dualism. Dreams, illusions, and hallucinations have shaped the contemporary notion of perception. Allegedly, the empirical evidence offered by cases of dreams and hallucinations is consistent with two widespread prejudices—namely, the mind is different from the world and the world is not what it appears to be. Can we upturn the applecart by envisaging a model of perception that comprehends all kind of experience? Can we explain hallucinations in terms of perception rather than the other way around?

So far, I have dealt with only the most favorable case—namely, successful everyday standard perception. In such a situation—the Shangri-La scenario—whenever one perceives a red apple, a red apple exists. Two germane notions—the spread mind and the spread object—have been deployed to show that one's experience is indeed identical with the object, at least in such favorable conditions. Can we extend this optimistic model to less encouraging cases such as hallucinations and the like? I believe it is indeed possible.

The explanatory direction can be reversed by modelling hallucinations as a case of perception—i.e., perceiving a real physical object. To do so, I will put forward a model of perception that can be extended to cases such as memory, illusion, hallucination, and perceptual error. From here onwards, as regards memory I will refer to episodic memory with phenomenal experience as when one tries to remember the color of one's sofa with the purpose of matching it with a new carpet. I will show that all such cases are cases of perception insofar as an actual physical object is available for each of them. Applying the model of the spread object/mind, the key idea is that *every "phenomenal" experience is the perception of a real physical actual object. Thus, all experience is perception and, in turn, all perception is identity with objects.* Crucially, I will show that such an account does not conflict with available empirical evidence.

Popular notions such as those of "vivid images playing behind one's eyes" or "inside one's mind" will be shown to be both empirically unnecessary and ontologically obnoxious. Locating consciousness in the objects outside our bodies seems strange at first, but in the end it does not introduce any conflict with our everyday experience and provides a long-sought ontological economy.

The aim of this chapter is to unfold the causal geometry[32] of experience and to show that—contrary to widespread popular notions but consistently with our experience—all cases from memory to hallucination are based on identity with actual physical objects. Here is a rough sketch: in standard perception, I see a red apple and the red apple is there. I see the red apple not because of *a relation* between *my mind, my brain,* or *myself* and the red apple, but because the apple, which is the actual cause of a process, takes place in the proximal spatiotemporal surroundings of my body. In all other cases—such as hallucinations, dreams, and memory—objects take place in an extended neighborhood of my body. The apple may be separate from my body by a few hundreds of seconds or by hours, days, weeks. It does not make any difference. There is always a finite time span. Historically, common sense named perception the experience of temporally close objects and memory/hallucination/dream the experience of temporally faraway objects.

The proposed solution is not farfetched because it does not suggest anything different from what happens in standard perception. Consider again the red apple, which I suggest is my conscious experience. The apple takes place sometime before the occurrence of any effect in my brain. Crucially, that apple exists because of the effects it produces in my brain—as in the fridge-light model of the object. A cause does not exist until and unless its effects occur somewhere at some later time. Likewise, my experience does not take place where my body is— in this case, it takes place on the table in the case of the red apple (Figure 5, top). My experience is not in my head. Of course, my body is not on the table but this is not an issue since my experience is the external object. Thus, in standard perception, the object is spatially and temporally located elsewhere with respect to the place and the time in which neural activity completes the causal process. However, since the distance is relatively short and the time is negligible if compared with human reaction times, such spatiotemporal aspects are usually ignored. In practice, though, nobody denies that, in everyday perception, a temporal and a spatial chasm separates objects and the ensuing neural activity.[33] I ask you to consider generalizing this fact to encompass all cases of experiences.

Do not forget that the main hypothesis is that *one's experience of the red apple is the red apple*. In standard perception, one's experience is an object that took place in some external location at an earlier time. The time span is never null. Experience is then identical with external objects taking place at a time and place different from where and when neural activity does. Such a hypothesis has a mandatory consequence: *for every experience one has, somewhere and at some earlier time there must be a corresponding physical object*. When we perceive, dream, or hallucinate, our experience will always be identical with an object or a combination of objects physically distinct from our body and yet physically occurring at some point in space and time. So the object we perceive is constituted by a combination of otherwise separate physical stuff. Sometimes, such an object is an unusually spatiotemporal object, but it is no less physical. As we will see, *standard objects are not any better, they just require less causal work*.

The crucial point is that whenever one experiences something—whether a perception, a dream, a hallucination, an illusion, a vivid daydream, or the upshot of direct brain stimulation—one does not experience something *inside* the mind, let alone inside one's brain. Rather, one is identical with an object that has the very features one experiences. I will show that such an object is as physical as the familiar red apple on the table. In this way, we no longer need to pull experience out of the brain like a white rabbit out of a magician's hat. The external physical world is all we need. In sum, the present account bravely faces the apparently impossible challenge of locating a real actual physical object for every case of experience. Consequently, if one perceives or dreams of a red apple, a red apple will exist at some space-time location. Such a red apple is one and the same with one's experience. Even more surprisingly, if one hallucinates a pink elephant, a pink elephant will have existed somewhere and at a certain time. Key to the proposal are the causal conditions for both perception and misperception. They differ only in details such as the temporal lengths of involved causal processes. All experience is identity with physical objects.

· · ·

If I experience something and everything is physical, something physical must be my experience: it must have the properties of my experience. I don't beat around the bush—if I experience a flying pink elephant, a flying pink elephant will exist somewhere and will be the cause of my current brain activity! Experience and neural activity cannot be the same physical entity because, trivially, they have different properties. In Shangri-La, all experience is perception of objects that happen to be in front of the beholder. Thus, objects are plausible physical candidates. It's easy to know where objects are. However, we do not live in Shangri-La and thus we must face misperception, illusions, dreams, and hallucinations. Can we stretch perception so that it encompasses such cases? Can we single out a real physical object every time we dream or hallucinate? Here, I will show that it is not only possible but also much easier than usually assumed.

Consider a familiar case, our beloved red apple. I sit in front of the red apple. Between my eyes and the red apple, there are roughly three feet. Light rays take approximately 1×10^{-12} seconds to travel from the apple's surface to my eyes. Once they reach the retina, the causal process needs an additional 100-300 milliseconds to complete the neural activity that allows the apple to be recognized as such.[34] In Figure 5, the object "red apple" takes place at a certain time and distance from where and when its neural effects occur, which are not my experience. As usual, the key hypothesis is that experience is identical with the object rather than with the neural activity. Experience is *where* and *when* the red apple takes place. The advantages of such a relocation begin to be manifest. Revealingly, nobody holds a punctual view of the physical underpinnings of the mind which would be physically implausible. Even the most hard-core neuro-centric scholar takes into consideration a set of neural events spread in space and time. Any reasonable amount of neural activity spreads over a non-punctual spatiotemporal area. Neural activity itself is spread across both space and time. A neural activity that spans 300 milliseconds requires 30 meters of neural connections to complete. The finite time required by neural activity is key. In fact, if neural activity is spread in space-time, such a space-time spreading cannot be a reason of concern and can be taken into serious consideration, in a much larger scale.

Like neural activity, which is spread in space and time, the perceived object is spread in space and in time. Sometimes, the temporal extension of the object is obvious as when we listen to a piece of music. At other times, it is less obvious as when we watch a gesture or a still object. When we perceive a movement, we perceive something spread in time. A gesture takes time to complete. In vision, the apparent lack of temporal extension is the offshoot of time delays so fast as to be subliminal. Yet, different visual features—e.g., color vs. shape—require different times in order to be perceived. Consequently they originate in different moments in time.[35] For instance, a face is an object made of several features spread in time. Colors, expressions, and shapes originate in slightly different instants of time. Everyday standard objects are spread both in space and in time.

Consider again the external object and the end of the neural activity (Figure 5). Their locations are different both in space and in time. The location of neural activity is vague and fuzzy, but it can be approximately envisaged. In the absence of any natural threshold, *the spatiotemporal gap between these two locations is arbitrary. It is fixed only by causal processes.* Of course, the proper kind of physical connection must hold between them. The basic idea is that the object takes place thanks to a causal process whose completion is possible because of the neural activity. However, such a causal relation between the object and the brain does not entail that the experience takes place inside the brain. One's experience is identical with the external object, no matter where and when such an object is. Moreover, the existence of the object is possible because of the corresponding neural activity—experience and the object are the same reality seen from different angles.

How much can we stretch the space and time between the object and the neural outcome? Since there are no natural thresholds, the answer is . . . as much as we like. Why should a difference in the length of the time span result in a change in the nature of the phenomenon? Consider what happens when you look at the moon. Is looking at the

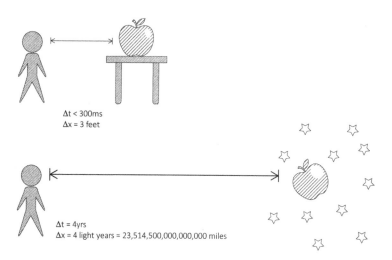

Δt < 300ms
Δx = 3 feet

Δt = 4yrs
Δx = 4 light years = 23,514,500,000,000,000 miles

Figure 5. The spatiotemporal distance between the object and neural activity can be arbitrarily great.

moon not a case of perception because of the time—give or take, one second—that moonlight takes to reach your eyes? Looking at the moon is like looking at an apple on the table. What about the sun, then? Sunlight takes eight minutes to reach us. What about the stars? Is looking at the stars different from looking at the red apple on my table? The time gap can be as great as thousands of years or more. The farthest object visible by the naked eye is the Andromeda Galaxy, which is a staggering 2.5 million light years from here. They are all cases of perception. I perceive the stars—and the sun and the moon—just as I perceive the red apple. Any model that applies to the latter must apply to the former.

In principle then, the time delay between the external object—the red apple—and the end of the causal process can be arbitrarily long. For instance, move the red apple farther and farther away—first a few feet, then thousands, and then up to billions of miles. If the red apple were bright enough and your sight powerful enough, you would still see it (Figure 5, bottom). Because of the longer space and time between you and the red apple, would your experience cease to be a perception? No, of course not. The increased distance determines a longer causal process in both space and time, but the nature of the phenomenon remains the same. Moreover, neither spatial nor temporal magic thresholds are crossed. Putting aside practical obstacles, we can imagine increasing the temporal delay as much as we would like. *The amount of space and time is not and cannot be a key factor in distinguishing perception from other forms of experience.*

A handful of philosophers have noted the likeness between perceiving a nearby object and looking at a distant star.[36] Roderick Chisholm argued in his book *Perceiving* (1957) that we cannot contrast the perception of everyday objects with the perception of distant stars solely on the basis of the length of light travelling:

> [...] the paradox arises only because we tend to assume, until we are taught otherwise, that any event or state of affairs we perceive must exist or occur simultaneously with our perception of it. [...] The perception of a star that is now extinct should be no more paradoxical

than the action of such a star on a photographic plate or its reflection in the water.

By the same token, occasionally A. J. Ayer considered whether seeing distant stars suggests that we have direct perceptual contact with the past, and that memory and perception are more akin than they have so far seemed.[37] For one, the philosopher Yuval Dolev believes there is a difference between perceiving astronomic objects and earthly ones.[38] If we link the idea of an arbitrary timespan with the notion that perceiving is being identical with an object, we have a solution—in one's past, which is still causally present, an object is always available.

We are back in Shangri-La.[38] One's experience is the object that has the properties of one's experience—*no matter where and when the object took place.*

When you experience a red apple in everyday life, your experience is the red apple that takes place as the actual cause of your neural activity. The causal relation does not snap back, dragging the form of an external object inside the mind, like a hook with a fish. *Representations, presentations, acquaintances, and intentional relations can be set aside.* The brain does not have to create an isomorphic copy of the world. An object is on the table with certain properties, and such an object is one's experience. Our experience is there too. For the same reason, it does not matter where and when the red apple took place. The object's existence is actual because of the occurrence of an effect. The causal process stretches across arbitrarily large spans of space and time.

Experience is spread since the actual object can be arbitrarily far away in both space and time. Causal processes stretch across possibly huge space and time spans. Why should an apple be a better candidate for my experience than a faraway constellation? As with the red apple, does the constellation also exist because of one's brain? Surprisingly, yes. The constellation that, say, Emily sees in Boston on the night of July 13, 2013, which is not the constellation other inhabitants of the galaxy might see—on other planets and on other nights—does not exist until the light rays emitted by the stars reach Emily's eyes. Like the stuff of which an apple is made, such stars might have already been many different things. They might have been composed of different formations. Yet, they cannot do everything by themselves. They need to interact with other physical systems to pull into existence all that they might have been. Surely, some of the things they might have been, have already happened. However, because of their position, mass, temperature, chemical composition, and brightness, such stars have the capability to extend their causal influence across intergalactic distances. Due to nomological constraints such as the limit of the speed of light, what the stars might have been has to wait millions of years before getting into Emily's retina and producing an effect. Yet, the constellation is not complete until it runs into Emily's eyes. When the process gets to an end, it pulls into existence its cause, namely the constellation. Like everything else, the constellation has a relative existence. *After the effect occurs, but only afterwards, the constellation has always been there.*

We can discuss forever whether the constellation existed before producing an effect inside Emily's brain. The discussion is of no interest. For Emily, no constellation existed before the stars produced an effect in her brain thereby fixing its existence. Similarly, Emily's constellation does not exist for anyone else. I do not refer to the light of the constellation, but to the very spatiotemporally-scattered constellation. One is inclined to assume that the constellation has had gazillions of occasions to take place before having causal intercourse with Emily's body. While this case is very probable, it is not necessarily so. Such a fact does not contradict the present account. The theory of spread mind requires a physical object for any experience. Of course, objects can occur without partaking in anyone's world. An apple is always an apple whether it is part of the subset of the world that is one's experience or not.

A mandatory caveat. The theory of spread mind neither endorses nor requires any form of backwards causation—in other words, causation going backward in time. Emily's perception does not have a temporally backward, retroactive effect. Only forward causation is invoked. In the present the stars exist but are not yet a whole. The constellation is, so to speak, ontologically vague until, at some ensuing time, those stars will produce a joint effect. In this case, Emily's body, which is just an object, will provide the required enabling causal circumstances. In this case, the stars seek having causal intercourse with Emily's body to breed a constellation. The present is undefined until its causal effects come to an end. All of them. Interestingly, in the case of the photons going from a star to one's eyes, relativity theory backs up this view literally. Due to time dilation, the internal time of a photon does not tick. Thus, for a photon, no time goes by on its way from a star to Emily's retina. The presence of the photon, then, is fixed by the end of its journey.

∘ ∘ ∘

To convince yourself that the spread mind view has the resources to reduce all forms of experience to perception, a simple optical device that mimics all known cases of experience such as dreams, illusions, and hallucinations in terms of direct perception is used as an intuition pump. The model will achieve such a result by tinkering with the external geometry of optical phenomena. Using an optical device to model various forms of experience and perception has many advantages. First, if the attempt is successful, an optical device will show that all cases of experience can be reproduced by means of changes in the external geometry of perception. Second, optics does not require notions such as information, representations, mental images, and the like. Of course, the above is only a useful analogy, I do not claim that the brain exploits optical causal chains. However, the brain exploits other causal chains by means of sense organs and neural networks. In the end, both optical and neural processes are just causal chains.

The ideal device must satisfy three requirements. It has to be transparent. It must be able to introduce an arbitrary time delay. It should behave like a false mirror: namely, it must be able to merge two light rays into one. Such a device is technically feasible.

So, back to Emily and her red apple. Consider a transparent medium—e.g., a transparent glass between the red apple and Emily's body. The glass is so transparent that we take the trouble to stick a yellow post-it note on it to alert people of the possible danger (Figure 6, left). Emily looks at the apple through the glass. Does the apple look any different? Of course not. Is there an image on the transparent glass? No, there is not. Is this situation any different from perception? I don't see how, nothing has changed. The glass is tantamount to the air between one's eyes and the objects. Space and time have not changed significantly. By the same token, when we look at the world through a window we perceive the world. Likewise, people who wear glasses perceive the world, too. Yet, the glass has introduced both an unperceivable delay and a physical medium. Nevertheless, such a delay does not change the structure of the causal process.

Next, we increase the delay—i.e., the time span between the external object and the neural activity. Between Emily and the apple, now, there is the transparent glass. Unbeknownst to Emily, a mischievous scientist has added an invisible device that slows down the light as it travels across the glass (Figure 6, right). Amazingly, this is technically possible today. While it is impossible to go faster than light, certain devices can slow down light.

Suppose then that it is possible to set an arbitrary delay. We start by small amounts. At first, the delay is barely perceptible. Is it still a case of perception? Why shouldn't it be? A delay does not rule out perceptual processes. As we have seen, due to nomological constraints such as the speed limit of light and average neural signals speed, even in the most favorable cases of perception a delay always separates objects from neural activity. There are many ways to increase the delay. For instance, moving the object far from the beholder will increase the time delay, as is the case with planets, stars, and galaxies. In other words, *what we see is always in the past, yet in a past that is our present.*

Figure 6. A transparent medium does not change the nature of perception (left). Neither does a time delay (right).

The conclusion is that a time delay—due to the glass or the distance—does not rule out perception. If one increases the time delay gradually, perception becomes uncomfortable, yet the causal structure of perception remains the same. Emily continues to perceive a red apple no matter how much time light takes to get to her eyes. Human perception is only apparently instantaneous.

Then we turn the glass into a false mirror capable of merging multiple light rays. Most transparent glass acts both as a reflecting surface and a transparent medium (Figure 7). Now the optical device combines two light rays into one. Such a common phenomenon is often a cause of discomfort for laptop users and a minor amusement when one peers inside shop windows. When this happens, one sees a combination of real objects. Such a combination is a composite object with actual causal power. The unusual geometry of light rays gives causal power to this composite object. Once again, though, the different arrangements between what one sees and what lies in one's surroundings do not change the nature of perception. Such changes in the external geometry of vision do not imply that what one perceives is less real. One perceives a physical object, albeit an object that is spread in space and time—i.e., a spatiotemporally composite object. The causal continuity with external objects is alike to that occurring in standard perception. Because of the partial transparency of the screen, multiple causal pathways single out an unusually spatiotemporally distributed object.[39] Crucially, one does not see a mental image, but a physical object akin to the red apple.

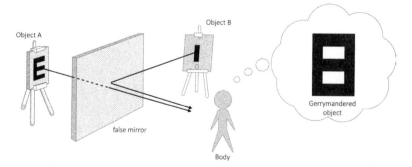

Figure 7. Multiple causal pathways allow spatiotemporally composite objects to be actual causes.

This is as good a place as any to dispel a common misconception about mirrors that might be a source of confusion—there are no images on mirrors: *No inverted image hovers on the mirror. Rather, a mirror changes the normal geometry of light pathways.* In fact, through mirrors we see the real world—such as we do while we drive and take advantage of rear-view mirrors. Watching the world through a mirror is akin to watching it through a glass. In both cases, a continuous causal chain obtains between external objects and one's eyes. However, in the case of mirrors, there is a catch: light rays do not follow a single rectilinear trajectory but instead follow a trajectory composed of two segments. In everyday perception, light rays are rectilinear. In mirror-mediated perception, a light ray is composed of two segments. Since the situation is relatively unusual, mirrors are often (mis)conceived as objects having a left-right inverted image hovering on their surface. This is wrong. No image hovers on a mirror's surface. The idea that there is an image over the mirror is yet another example of the fallacy of the intermediate entity; an issue I tackled in a philosophical cartoon in 2010. The mirror is an object that changes the causal geometry of incoming light rays. Thus, if I look at the red apple by means of a mirror, I can informally say I see a reflection of the red apple, but it is just a manner of speaking. I do not see a reflection, aka an image: I see the real red apple by means of unusual pathways. There is no such a thing as a reflection. I do not see reflections. I see objects.

In conclusion, we have at our disposal technically feasible optical devices that may be placed between one's body and any object one encounters during one's life—i.e., a half-transparent mirror with a controllable delay. Such a device allows one to perceive any moment in one's life and, even, any moment before one's birth. Of course, one should have positioned the device at the right time and place. Otherwise, one couldn't use the device later. Yet, if one had positioned the time-delaying false mirror at the right place and at the right time, one could see what happened at any time later. This is crucial. In fact, our brain seems to be unable to access any elementary phenomenal experience unless it has had direct contact with it during the lifespan of our brain.

Using such a device, one might see past events as though they were part of the present. More precisely, due to such a device, past events would still be part of the present. All these events, and their combinations, are then as much part of one's present as the red apple on my table now. The key idea is that such a causal geometry is the causal structure that underpins memories, dreams, and hallucination. If the requirement of placing the device at the right time and place seems unfeasible, practical solutions can be envisaged. For instance, one might place millions of such devices during one's life, a multiplicity of such mirrors with multiple time delays. In such a way, in the future, one will always have one mirror placed at the right previous time. Call it the prodigal strategy.

In fact, this is exactly what the brain does in causal and neural terms. In the example, using such an optical contraption, one does not *imagine* or *remember* a past event, one *perceives* it. Dreaming is not like seeing a picture, as if watching a television screen. Rather, one sees the original object as though one had it in front of oneself, only that the causal chain is longer. The perceptual process is alike to that which links our bodies with a constellation, a star, the sun, the moon, a mountain, a house, and—finally—our beloved apple.

.   .   .

The envisaged optical device mimics not only perception but also all forms of experience. Using the right optical tools, and applying the prodigal strategy, *any object with which we have been in contact during our life, at any time and at any location, will still be present in the same way as the red apple in front of us is present. Furthermore, thanks to false mirrors, any combination of objects and properties will exist as a composite yet physical object.* Does such a model match all cases of experience, from memory to hallucinations? I bet it does.

Traditionally, memory has represented a common objection to a purely perceptual model of experience. Many people have the feeling that their memories are stored somewhere inside them, like a repository of videotapes, pictures, or images—whatever metaphor current technology suggests. In his popular novel *Kafka on the Shore* (2002), Haruki Murakami writes

> Inside our heads [...] there's a little room where we store memories. A room like the stacks in this library. And to understand the workings of our own heart we have to keep on making new reference cards. We have to dust things off every once in awhile, let in fresh air, change the water in the flower vases. In other words, you'll live forever in your own private library.

Yet, nothing in our experience proves that what we recall is inside our brains. Our memories are not sticky with neural goo, so to speak. Though we are able to recall our past, nothing dictates that we must have a physical copy of our past stored inside our heads. What we need is a mechanism that allows us to experience our life episodes—a mechanism that allows past events to be present. Projecting neural records against a mental screen is not such a mechanism: that would be more like a magic spell. Rather we need a mechanism that marshals still-causally-relevant events as part of the present. *We literally see the past.* Memory is a form of temporally extended perception. *What we call past is still present.*

Consider the simplest case, veridical memory, which is memory of something that has really happened. Our bodies are proxies that extend the causal influence of events and objects. I ask you to remember the color of your sofa without looking at it. As a result, you recall the last time you saw the sofa. By doing so, to some extent, you see your sofa and its colors. How is it possible? The traditional explanation is that you see it "in your mind" by means of some vicarious mental image that no one else can see. Yet, your mind is not a physical place where you might step inside to have a look. The theory of the spread mind offers a different solution. You can still see your sofa where and when it was when you saw it, say, this morning when you sat on it. You see your sofa in the same way you see Betelgeuse or the red apple—namely, because a causal chain links the object you experience, which is your experience, and your brain. All three phenomena—the star, the apple, and your sofa—are causes that take place thanks to the structures in your brain. When you remember something, you do so by means of some neural structure that plays the same role as the optical device just described. Body and brain are causal proxies for external past objects and events. The past is still causally effective by means of your body. The past is as present as the events and the objects one customarily places in the present.

If one had adopted the prodigal strategy, an optical device with the right time delay would be available for each event in one's life. Each optical device would be committed to a particular event and that event would wait to be the cause of a process. In brains, the causal processes of perception are slowed down, not by means of optical devices, but by means of neural processes. *Memory is delayed causal influence implemented through neural structures that allow past events to exert their influence after an arbitrary span of time.* After all, brains are causal proxies for objects and events spread in space and time. When I remember a past event, such an event exerts an effect at the time that I recall it. Yet, the event remains where it was.

Remembering is akin to placing a temporally-delayed glass at the time of the event to be remembered. As argued above, such a device would do the trick of making one's past present and thus visible. After all, in everyday perception, light rays reach one's eyes after an arbitrary span of time. Once again, such a mediated perception is standard perception and a model of veridical memory. Given the appropriate device at the right time and place, observers could see any event of their past life at any later time. In brains, neurons extend the causal influence of events ahead in time. In this way, past events and objects are still part of one's present. *Remembering past episodes in one's life is akin to seeing distant stars.* In this regard, in 2007, Yuval Dolev expressed a similar view in *Time and Realism* (italics mine):

> Visual recollection differs, of course, from eyesight, most evidently in that one's eyes do not figure in it. But, although this difference is indeed blatant, it is also the only significant one. [...] *It is not the case that one is "direct" in a way the other is not. Memory is a faculty that broadens our field of vision to include objects and events that (at a given moment) are not visible to our eyes.*

Each receptor in the retina is causally linked with a light ray and thus with its source. In turn, many cortical areas are causally linked with past events and objects. The brain itself is a sort of super-eye where object-receptors play the role of photoreceptors—they keep past objects and events in existence, thereby accessing a larger present. The brain is causally analogous to a giant array of such causal units, each one devoted to a given object or event.

Suppose you have a huge array of devices, each with a different time delay. Each *device* is the causal gate through which an object in one's life produces effects. At any time, one accesses—and sees— what went on at a given moment of one's life. Each device singles out a given object at a given time. Each device is like a window on a past moment. The model requires neither stored information nor inter-mediate images to project on a mental screen. Identity with multiple moments in space-time enables the perception of one's past—i.e., di-rect perception. The array of devices does not store images. Each device acts like a mirror that allows the propagation of light rays. Imagine a mosaic of billions of mirrors, each with its own time delay, like a humongous wall plastered with windows mostly covered by shutters. The windows that are not shut let the light flow. One can choose which windows open. The open windows single out which set of objects or events are still causally present. Subjects perceive past moments of their lives, literally. They do not perceive memories, re-productions, replicas, representations, stored images, recollections, or engrams. Remembering is perceiving the past insofar as the past is still . . . present. The past is present because it is still causally active now. Perceiving is being.

Such a model of perception is akin to looking at the night sky. The field of vision stretches causally, spatially, and temporally. Each star belongs to a different moment in time, yet such stars fit together into constellations despite their temporal inhomogeneity. Likewise, by switching on and off neural structures, we select what place and time we see by choosing which part of one's past is still present.

· · ·

But now here is a worry: what about those cases—such as false mem-ories, dreams, and hallucinations—that seem to reshuffle one's past as freely as a synthetic cubist painter? Doesn't the proposal crumble because of cases such as the traditional flying pink elephant? Do not worry. The optical device we have just envisaged is capable of coping with such cases. It is time to take advantage of its third feature, namely, the merging of multiple light rays.

As long as one's experience is a combination of real experiences, it is possible to show that what one perceives is a physical object made of reshuffled physical parts, a spatiotemporal fruit salad no less real than a whole fruit. Dreams and hallucinations are free from the bonds of real life to the extent that they freely reshuffle one's past. One can dream of a mouse with elephant ears and insect-like legs. Yet, one must have had previous contact with mice, elephants, and insects. The building blocks of dreams and hallucinations are always traceable back to the real world.

Consider a kaleidoscope—a set of mirrors, sometimes totally opaque and sometimes partially transparent, that intermingles light rays in unusual ways, thereby causing one to see unusual combinations of whatever lies in front of the device, usually an assortment of brightly-colored glass fragments. As a result, when one looks inside the kaleidoscope, one believes to see a multicolored, symmetrical image. Yet, contrary to common sense, one does not see an image but rather the very fragments of glass through multiple pathways. With respect to everyday perception, the pathways are so complex that it is easier to assume that there is a geometrical image that has been created. Yet, there is no image inside the kaleidoscope. Inside the device, there are only mirrors altering the causal geometry of processes. The example of the kaleidoscope shows that, even in a domestic environment without any time-delaying capability, something akin to a geometrical hallucination can be obtained. Thanks to the kaleidoscope, an object takes place because of the changes in the causal structure of the environment.

Such a capacity to rearrange reality is well known with respect to glass and mirrors. In everyday life, e.g., in shop windows, I can see my head on top of an expensive suit without having to wear it. Normally this does not happen because shop owners try to minimize potentially distracting optical effects, but occasionally it does. Thus, one sees exotic combinations. Because of the geometry of light, such combinations are physical objects with physical effects.

Devising a spatiotemporal kaleidoscope, whose internal mirrors are akin to the abovementioned optical devices, is not a farfetched idea. While off the shelf kaleidoscopes have a fixed geometry and are not capable of time delays, one could implement time-delayed mirrors with a reconfigurable geometry. In this way, thanks to time delays and the combinatory power of false mirrors, merging arbitrarily spatiotemporally scattered objects is conceivable. It is possible to envisage a honeycomb of such units, which will be able to allow multiple paths to merge and mix in different proportions. Such a contraption will not contain any stored information. Yet, every time one looks inside the machine, one will see a past moment—by and large, *in memory one literally sees one's past.* The machine will slow down and combine the causal processes between brain and world. By selecting causal processes, the machine will bring into existence new objects. In this way, any conceivable combinations of past objects and properties can acquire causal efficacy, thereby being causally efficacious and, in turn, being real.

Dreams and hallucinations are explained in a similar way. They are situations in which one's experience *is* an object brought into existence by merging unusual causal pathways. Scattered objects—from any time and place in one's life—might suddenly produce an effect now. For instance, the dream of a pink elephant is a case of direct perception of a composite object made of pink patches, an elephant, and something flying. The three building blocks took place in three different space-time locations during one's life. However, due to the optical machinery just described, their causal pathways combine and give rise to a flying pink elephant. The resulting flying pink elephant is not a concocted mental image popping out from nowhere, but it is a physical object made of physical parts spread in space and time. Everyday familiar objects are not different in this respect. They too are made of parts scattered in space and time.

In our brain, neural pathways play the same role as the mentioned optical time-delay devices. When the constraints of everyday perception are weaker, the brain acts as a kaleidoscope. The brain is the condition for the causal efficacy of one's objects. When brain activity is causally detached from the proximal environment—as it happens, say, in dreams and sensory deprivation—temporally-more-remote causes find their way in. *Past objects are the stuff dreams and hallucinations are made of.* Such composite objects are just as real as everyday objects. The only difference is that while the red apple on the table is easily repeatable given many enabling factors, a dream is an object constituted by wildly scattered events. However, claiming that a flying pink elephant is not real is tantamount to claiming that the Ursa Major constellation is not real because the seven stars composing it are not where and when we see them to be. In fact, due to the limit of light, those seven stars have long since changed their position. The Ursa Major can be seen only in a limited region of space-time inside which the earth is currently located. However, when we see these stars, we do not see an image. We see physical objects. The Great Canary Telescope does not take pictures of images: it takes pictures of astronomical objects. When we stare at the night sky, we perceive stars and constellations just as we perceive the red apple. We are not looking at a mental image of the stars, or at non-existent objects. We look at real constellations in the sky. Ursa Major is as real as Mount Elbert.

A final objection to this optical model is represented by a myth in which many have placed their faith—that the mind is capable of creating any kind of content arbitrarily. According to this belief, congenitally blind subjects are allegedly able to dream or to imagine colors. Solitary thinkers are supposed to be able to concoct all kinds of phenomenal experience inside their minds. Similarly, dreams are supposed to produce all kinds of arbitrary sensations. One's inner world is supposed to be as rich as one wants it to be, no matter the physical world.

Reality is a far cry from this belief. Remarkably, to the best of our knowledge, in all cases of congenital absence of a sensory modality, subjects have lacked that particular phenomenal experience.[39] The only evidence about phenomenal experience independent of actual perception is a handful of cases, some of which are in dire need of further investigation.[40] Coherently, dreams and visual imagery are so constrained by one's perception that substantial changes in perceptual content trigger changes in dreams and mental imagery.[41]

A popular example is offered by phosphenes. It is well known that direct stimulation of the brain (by electric, magnetic, or mechanical means) can trigger elementary visual experiences called phosphenes.[42] This fact is customarily taken as evidence that visual experience is generated in the brain without need of actual light, for instance by the philosopher Ned Block in his rebuttal of externalism.[43] But this is a hasty conclusion. The available evidence would be conclusive only if completely congenitally blind subjects had visual phosphenes—i.e., subjects whose individual history has never included any light. As a matter of fact, however, no congenitally blind subject has ever reported any phosphenes.[44] Once again, a convenient myth has backed the internalist notion of the mind.

Frequent exposure to a certain class of stimuli constrains the phenomenal repertoire of dreams. For instance, during the 1950s, the widespread consumption of black and white images led to a sharp rise in the percentage of black and white dreams, to the extent that some psychologists questioned whether we ever dreamed in color at all.[45] Surprisingly, some authors draw the opposite conclusion—namely that subjective reports about mental activity are incorrect.[46] Such a conclusion seems to imply that the phenomenal experience during dreams remains the same no matter what one perceives during the day. Since it is assumed that experience has to be phenomenally-colored, many psychologists consider such cases as evidence about the unreliability of introspective reports. Yet, if we look at the evidence from a different perspective, we do not need to question the reliability about personal experience. Dreams are either colored or monochromatic because they are a postponed perception of one's environment. If people spend a considerable amount of time watching monochromatic pictures and movies, their dreams will be made of achromatic objects. One will have colored dreams of black and white objects rather than black and white dreams of colored objects. Of course, such an alternative account does not deny that a standard trichromat cannot dream in black and white. In such a case, one will dream of a subset of the actual world.

· · ·

As of Descartes's time, perception has been modelled in causal terms. Roughly, this model suggests that an external object causes a brain activity, which, in turn, *causes* an experience. The causal argument is apparently solid but is built on top of metaphysical and empirical premises that are questionable. Such premises, although cherished by tradition, are not empirically conclusive. For one, experience is assumed to be temporally synchronous with neural activity. In other words, many have assumed that there has to be a special neural activity that, whenever it happens, is followed by conscious experience. This has never been empirically tested or conceptually clarified. If experience is both physical and different from the external object, how could it be related with an object? Neural activity and external objects are two separate physical entities. Otherwise, if experience is not physical, how could it relate to both a physical object and with neural activity?

Notwithstanding its many shortcomings, the causal model has remained very popular. Nicolas Malebranche expressed the causal argument in the following terms (italics mine):

> Let us suppose, too, that God imprinted upon *your brain the same traces, or rather produced in your mind the same ideas*, which we take to be present now. [...] supposing that the world were annihilated, and that God nevertheless produced *in our brains the same traces, or rather in our minds the same ideas*, which are produced in them on the presence of objects, we should still see the same beauty.

The point is clearly stated. The external world has a contingent causal role. It might and it might not be there. Physical objects produce traces in the brain. Such traces, in turn, produce certain ideas. If God wanted to skip the creation of external objects, God would produce only properly configured brains. Yet, why bother with brains? God could dispense with neural traces too. In a dualistic metaphysics, God could produce "in our minds the same ideas!" and that would be more than enough. Of course, this view works fine only if one is a dualist. If one is a physicalist, God, too, should tinker with one's physical underpinnings. If one adopts a brain-mind identity theory, God must take care of one's neural activity. Thus, unsurprisingly, the causal theory makes sense only if one supposes that the split between appearance and reality holds.

The crux of the matter is that we do not know whether the premises—the sufficiency of brain processes to produce experience and the separation between appearance and reality—are correct. For one, I believe they are wrong premises. Neither do we know how to make sense of them. One cannot assume the appearance vs. reality dichotomy and then use the causal argument to back it up—that would be begging the question. Worryingly, such premises are not conceptual truths but empirical hypotheses about the world.

In contrast with the standard view, the theory of spread mind suggests that the physical basis of experience is not brain activity but rather the external physical object—i.e., the object that is one's experience. The causal argument collapses because its key premises are set aside. Mind and world are no more connected by an awkward causal relation. Mind and world are bound by identity.

Many consider it a truism that the brain secretes consciousness. However, such a sufficiency of the brain is precisely what we do not know. We have neither empirical evidence nor theoretical arguments. Why should one opt for the brain rather than for the external object? Is there any difference between invoking a spirit and producing phenomenal experience out of neural activity?

The theory of spread mind offers an alternative stance, that there is no more need to proceed along the causal chain. In fact, according to the view presented here, *experience is the object one experiences*. The spread mind brings together identity theory and causal theory into one coherent picture: *causation holds between the object and the brain; identity holds between the object and experience*. The spread mind is a sort of union, a hybrid of causal theory and identity theory. The identity holds between experience and the object rather than between the brain and experience. The causal relation holds between the object and the brain rather than between the brain and experience.

The causal geometry of perception presented here suggests that one's experience is identical with the external object. Shangri-La paves the way for the identity between experience and object rather than causation. Therefore, the traditional causal chain is modified as follows.

$$(O \text{ is } E) \rightarrow N$$

Here, once again, one's experience is identical with the actual cause of the neural activity. The emphasis on considering an *actual* cause ensures avoiding all issues connected with dispositional and thus abstract, potential, in-actual, or in-existent entities. The causal process is not uninfluential though. On the contrary, it is key because it pulls the object into existence. Since the object is one's experience, in a sense, the causal argument has its final revenge. The object takes place because of the neural activity. Going back to the lake-dam metaphor, the dam is there because of the lake, but the lake exists because of the dam. One's brain and the body are the conditions for the existence of the external object that is one's experience.

But closer to the body is not necessarily closer to the mind. Why should it be any easier to explain how we perceive an "internal condition" rather than to explain how we perceive an "external object"? The notions of internal and external are misleading attributes! Why should acquaintance with the properties of internal bodily states be any more accessible to one's mind than, say, a red apple? Or a star gazillions of miles away? Or the gestures of one's grandmother made many years before? We can set aside parochial and obnoxious prejudices, such as the belief that to be inside one's mind something must also be inside one's body.

•   •   •

By means of the optical model of experience, it is conceivable that, for every experience, a corresponding external physical object with the very properties of experience exists. Such an object is a physical object at a certain spatiotemporal location. Furthermore, in contrast to an entrenched belief, no empirical evidence backs up the notion that brains can create a pristine mental experience—such as a new color one has never seen before. If this view is correct, all experience— whether dreams, illusions, or hallucinations—will be made of physical components that are the physical causes of what goes on inside one's body. Such causes produce a physical effect by means of one's neural structures. Therefore, such causes are the present no matter how much time they took place before neural activity. They are as present as the red apple on the table, only their causal connection is more articulate and extended. After all, *all experiences are spatiotemporally composite objects.* Reassuringly, though, all physical objects are, to various extents, spatiotemporally composite objects.

Having a physical object for every experience allows us to put forward a major attack on the traditional separation between subject and object, between experience and nature, and between appearance and reality. The price to pay—and the main risky prediction—is that pure mental experiences cannot occur. In this regard, I will pick up the gauntlet and say that for any experience, a physical object has to be available. Of course, one can always draw wrong inferences about one's experience. We might have misbeliefs about our world (and thus about experience). Marco Polo may be wrong about whether the animal he sees is a unicorn, but what he sees is still a part of nature. Obvious counterexamples such as illusions will be addressed in terms of misbeliefs about what one thinks one ought to see. *All perceptual errors can be revisited as erroneous beliefs.*

How do we perceive something? Because we are the objects we perceive. *Nothing is more intimate than the identity bred by the causal intercourse called perception.* In short:

- All experience is perception
- All perception is identity with a physical object
- Physical objects are actual causes of processes ending in one's brain
- Actual causes exist thanks to their effects
- One's mind is the set of such objects

This account is not an idle speculation but rather an empirical hypothesis that fosters empirical predictions amenable to falsification. The view predicts that, for each experience, a physical object must exist. One experiences something that is the actual case of some activity in the body. Every time I have an experience, an object with the same properties of my experience takes place. I contend that my experience and the object are one. Such an object takes place anytime earlier and, crucially, exists because of an effect inside my body. Without my body, no effect would have occurred and thus no cause would have existed either. Of course, the contingent relation between the external object and one's brain does not mean that my experience is an effect within my brain like, say, a neural activity in the fusiform gyrus. Rather, the effect brings into existence the object that nonetheless remains in the world where and when its elements were.

Once we have freed ourselves from the parochial prejudice that the mind is inside the body, our experience can be fleshed out in terms of collections of external objects. The theory of spread mind shows that world and experience indeed have the same properties.

In the actual world, nobody has ever seen anything like a perception, a representation, a relation, a phenomenal character, a self, or a soul. We have always seen only *objects taking place and causing other changes in other objects.* The red apple is an object. My brain is another object. My experience cannot be anything but an object. Which one? I suggest the red apple. Nature is made of objects all the way down. Nobody has ever seen anything but objects.

A theory of consciousness ought to show how to model nature in such a way that experience fits within it. The theory of spread mind is just such a theory. It suggests that all experience is perception and that perception is identity. If the view is empirically correct, identity is the only way in which the mind can be placed within nature. If nature is made of causally-related objects and experience is a part of nature, *experience must be made of objects too.* The main factor, which has prevented scholars from considering such an option, is the entrenched belief that appearance and reality may go askew and that subject and object are separate. These beliefs, in turn, originated from other beliefs of alleged empirical origin: that the mind can create arbitrary content regardless of the physical world and that experience can be different from the perceived object. In short, they amount to a cluster of arguments from hallucinations, illusions, and misperceptions. As I will try to show, such beliefs are empirically false. The present account shows the opposite—*experience cannot occur without corresponding physical objects*. The theory of spread mind shows where and when such an object—a spatiotemporally composite object—is located.

4.

Illusions

There are in fact no illusions of the senses, but
only mistakes in interpreting sensational data
as signs of things other than themselves.

—BERTRAND RUSSELL, 1948

THE BELIEF IN a separation between appearance and reality—between mind and world—has largely been fueled by cases in which allegedly one experiences something that is different from the alleged physical reality. It is a common truism that our minds can experience the world differently from what the world is.

Illusions are often presented as the paradigmatic condition in which things look different from what they are, thereby endorsing the separation between appearance and reality. For instance, one perceives a gray patch as though it was colored, or two lines of equal length as though one were shorter. Such a popular account is empirically wrong. While illusions are often presented as cases in which one perceives properties that do not match the physical world, they are a mismatch between the beliefs about what one ought to perceive, and what one actually perceives. If one had no expectations about what one ought to perceive, the notion of illusion would never apply. In short, *illusions are misbeliefs, not misperceptions*.

Traditionally, illusions refer to cases in which something appears different from what it is. In this regard, A. D. Smith wrote in 2002 that an illusion is "any perceptual situation in which a physical object is actually perceived, but in which that object perceptually appears other than it really is." Consequently, in philosophy, illusions are distinguished from hallucinations insofar, in the latter, that we perceive something that is not there at all.

In fact, most if not all accounts of illusions depend on something that is not always explicit, namely, *the nature of the alleged property one believes to perceive.* How do I know what the property is that one really perceives? I see, say, pink and I think that I should see white. Why should I see white? How can I rule out that, when I see pink, the property that is there, in the world, is not pink? After all, if I see pink, the most obvious hypothesis would be that the property I observe is, guess what . . . pink! How can I know what my experience ought to be? An illusory experience is not illusory per se, but happens because one compares what one actually perceives with what one expects to perceive. One cannot even conceive illusions unless one has some normative notion about what one should perceive. Illusions are the offshoot of the acceptance of an external authority—be it science or common sense—that defines what one ought to perceive. Experience versus knowledge.

Consider the following example. Emily has no *a priori* knowledge about colors. She has always seen a certain patch as having a certain property that her peers refer to as red. Eventually, in high school Emily is taught that red is indeed a certain range of light frequencies. After scrutiny, she finds out that such frequencies are not reflected by that patch. Emily is puzzled. On the one hand, she might insist that the patch and, say, the strawberries share the same property because she experiences both as having the same property. The patch is then red. On the other hand, Emily might feel the weight of scientific authority and bow to it by dismissing her own experience in favor of the officially authorized knowledge. In the latter case, the patch is not red. The mismatch between her experience and her acquired beliefs is solved by downgrading experience to mere appearance. The patch is not red (so science says!) but it looks as if it was so by means of being illusory red. Such a misleading notion of illusion suggests that the patch and the strawberries might have different properties. Yet, such a conclusion is

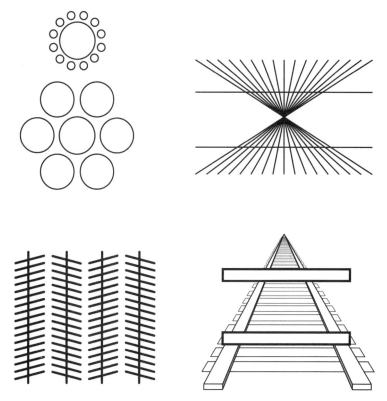

Figure 8. Classic visual illusions (clockwise from top left): Hermann Ebbinghaus (size), Ewald Hering (curvature), Mario Ponzo (size), Johann Karl Friedrich Zöllner (parallelism).

based on the unquestioned premise about what red *ought to be*. But why should the received model of color—be it light frequency or a more complex cluster of spectral properties—be any better than Emily's experience with respect to singling out a property from the physical continuum? Note that I am not suggesting that red is a subjective property. Emily's body, the patch, and the strawberries are all physical systems. Their combination and the fact that the patch and the strawberries have something in common do not require any appeal to mental subjective properties. Red is the physical property that causes effect via Emily's body when she sees strawberries, lips, Coke cans, and that patch.

The crux of the matter is the acceptance of a normative account of red supported by the authority of science. An illusion is not a perceptual phenomenon but rather a case in which one's experience, and one's beliefs about what one ought to perceive, clash.

In most cases, we presume to know what we should perceive. It is assumed that what one perceives has to be the absolute length, or the color that is reflected in normal conditions, or the shape that can be seen from a canonic viewpoint. Yet, in all such cases, the notion of illusory perception is based on the contrast between what one perceives and what is believed one ought to perceive. The latter is not an empirical fact, but rather it is an expectation derived from some pre-existing conceptual model of the external world.

In fact, a naïve perceiver has no way to tell whether a given perception is illusory. I look at a patch on a sheet of paper. I see it as pink. If I had no additional knowledge, why should I wonder whether the patch is pink? As far as my experience is concerned, I am not aware of any illusory aspect. The patch looks pink. If I look at, say, a Müller-Lyer picture, I see two lines that are different in some respect. Moreover, if I had no better knowledge about such lines, I would draw the conclusion that they are different. In fact, I will argue that, notwithstanding a common opinion, they are different. That's why they look different.

In sum, an illusion is a contrast between what one experiences and what one assumes one ought to experience. Illusions arise from a difference between one's experience and one's beliefs. However, beliefs can be wrong. Experiences cannot be wrong.

Illusions have to be explained in terms of actual experience, otherwise the Shangri-La scenario would fail. The theory of spread mind cannot survive any case of separation between appearance and reality since it is based on the identity between experience and physical objects. There has to be an object. No space is left for appearances that are not grounded in the physical world. Yet, such a responsibility is not a bad thing. On the contrary, it is a very good thing—it is a sign that *the theory of spread mind is a scientific theory that puts forward risky falsifiable predictions.* Empirical cases of illusions offer a first challenge and a good test for the spread mind. Is it possible to show, for every case of illusion, that there is an actual object and that such an object has the property one experiences? I believe this is the case.

.   .   .

If illusions can be explained away in terms of erroneous beliefs, why have scholars held the notion for so long that things can appear differently from what they are? The most likely culprit is that the standard view encourages being lazy: that an immaterial illusory experience is like an ontological joker. It always wins. If appearance and reality were separate, then whenever I experience an unexpected property, what I perceive might be illusory. *Assuming that our perceptions are wrong and our beliefs are true is all too easy.* The historical underpinnings of such an attitude can be traced back both to Plato and to Galileo, who tried to convince us that experience is of little value and that we must do violence to our senses. In a surprisingly similar manner, they both transferred epistemic authority and control to an elite of professional scholars who were authorized to state what reality really is. The community of savants established that the world is made of lengths, sizes, shapes, and light frequencies. As a result, most people assumed that the properties that one ought to perceive are lengths, sizes, shapes, light frequencies, and so forth. What did not conform to such an authorized list of properties was explained away as an illusory mental property devoid of ontological weight. The list is the outcome of centuries of efforts. Such a list has been revised and updated several times. For instance, the list of the *real* colors has been changed several times.[47] Furthermore, such normative models are only tentative, and often rather poor. They are only rough approximations of the actual phenomena we perceive through the senses. We perceive the physical phenomena our bodies single out from the environment by means of causal coupling. Such causal carvings are the offshoot of natural selection, individual development, and neural ontogenesis. No conceptual model matches exactly the properties our bodies pick out from the world. For one, consider how much the temperature we perceive is different from the neater notion of temperature as normally measured using standard thermometers.

In practice, though, *illusions arise because we mistake a property we believe we perceive for a property we actually perceive*—in illusions, one does not see an illusory property but a real physical property that is commonly misconceived. Most—if not all—illusions fit into the following two broad classes:

- One has wrong beliefs either about the property one perceives or about the property the object has
- One's body is altered in such a way that one perceives an unusual—but nonetheless real and physical—property

The former category (misbeliefs about what one ought to perceive) includes cases such as mirages, disguises of various sort, geometrical illusions, misbeliefs about intensity or hue or spatial resolution perception, as in the Checker-Shadow Illusion, relative size, closure, Enigma, Benham's top. We will survey a selection of them one by one. In all these cases, it is possible to show that one perceives a real physical property instantiated by the perceived object that, unusually, is different from the property one believes to perceive.

The latter category includes afterimages, attentional blindness, motion induced blindness, multi-stable figures, and—to some extent— Benham's top again. This category results from alterations in the physical structure of the subject. The subject perceives something unusual, which is nonetheless real and physical.

As it happens, the boundary between the two classes is fuzzy and partially overlapping. For instance, Benham's top belongs to both categories since it takes advantage of both mechanisms. This partial overlapping is not a reason of concern for the theory of spread mind. To get the gist of the argument, consider the traditional example of the room that feels warm to Emily who has spent the morning in the snow and feels chilly to Riccardo who has spent most of the morning sitting on his sofa. The room cannot be both warm and cold, therefore—or so common sense and the standard view suggest—Emily and Riccardo do not perceive the room temperature *objectively*. A philosopher can explain the case by suggesting that they perceive a *subjective* temperature—how the temperature appears to be is not what the temperature really is. *Real* temperature, on the contrary, is objective. In fact, any two thermometers, if not broken, will show the same temperature. Emily and Riccardo do not feel the same temperature. Thus, Riccardo and Emily fail to represent correctly the real temperature of the room. Right? Not at all.

The mistake offshoots from the assumption that Emily and Riccardo *ought to* perceive the temperature of the room as Daniel Gabriel Fahrenheit and Ole Christiansen Rømer defined it. Because of the authority of science, that particular model of temperature is taken to be the real phenomenon, while Emily's and Riccardo's temperatures are downgraded to mere subjective appearances. This is not necessarily the case. The property that Emily and Riccardo perceive is not the room temperature but some ongoing process inside the skin, which is, in normal conditions, sufficiently correlated with external temperature. It is a good match. Yet, occasionally, it might fail. Actually, it often does. Very roughly, the skin cells react to rapid skin temperature changes—e.g., touching either a warm cup of coffee or a cold piece of metal. Given the different conditions of Emily and Riccardo's bodies, what takes place inside their skin is physically different. Emily and Riccardo do not perceive Fahrenheit and Rømer's temperature but rather another physical phenomenon, a proxy property that human bodies use to estimate temperature. Thus, it is only right that they perceive different "temperatures." *Emily and Riccardo do not perceive the same phenomenon in two different subjective ways, rather they perceive two different phenomena, which are both as physical as Fahrenheit's temperature.* The coarseness of language is likely another factor that induces us to lump together many slightly different phenomena under the same name.

As a further proof, it is worth mentioning the fact that the property that Emily and Riccardo perceive is not Fahrenheit's temperature but is instead a certain chemical activity. Such an activity is caused both by differences in temperature between the skin and the environment and by molecules as diverse as capsaicin (spicy chili pepper molecule), allicin (garlic aroma molecule), allyl isothiocyanate (wasabi compound), and menthol. As a result, people perceive freshness when in contact with menthol and hotness when in contact with chili. This does not happen because, as is customarily said in textbooks, these substances *evoke subjective sensations*, but because the key physical process one perceives is the same both in the case of low temperature and in the case of menthol. The proof that the key process is similar in both cases is that they produce the same effects. Thus, one has the same experience because the phenomena one perceives are the same. The object one perceives is a composite object that comprehends both certain molecules and certain variations of temperature. One's experience is not mistaken. A cold iron bar and a menthol candy instantiate the same property and thereby produce the same effect. One perceives the world as the world is. Experience is not different from reality. We might have wrong beliefs about what we perceive and about what the world is. Once we correct our naïve beliefs about what we perceive, illusions are explained away. I claim that a piece of ice, a metal bar, and a molecule of menthol have the same physical property because they produce the same effect given a human body. Consider three keys that, although different in many respects, can open the same lock. If they unlock the same lock, no matter their overall features, they will share some fundamental physical aspect that is the cause of the unlocking. For the lock, the three keys exist insofar as they have the right causal structure.

Whenever we experience something, a physical property, which is the actual property we experience, occurs even in illusions. The mismatch between appearance and reality is a conceptual mistake rather than a perceptual fact. When one perceives the world, one's perception is unambiguously identical with the world one perceives. *One's beliefs about what one perceives might be wrong—one's perception cannot be wrong. According to the theory of spread mind, every experience is a physical object and cannot be wrong—it can only exist. Our beliefs might be wrong, but not experiences.*

The notion of illusion has gained momentum because of various factors: the assumption that appearance and reality are different, the application of an oversimplified model of the object, and the acceptance of normative notions such as beliefs about what one ought to perceive. To compensate for the mismatch between the properties one experiences and the properties one *ought to* experience, a fictitious

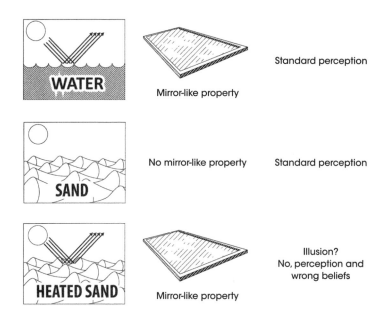

Figure 9. A mirage is the perception of a physical property that occurs both in the case of a pool of water and of heated sand.

mental image (*appearance* or *illusion*) has been introduced. I suggest the opposite strategy. During illusions, one singles out physical properties that are usually masked by other conditions. Illusions allow us to peer into the structure of our perceptual processes and to single out the actual physical properties that we perceive: actual properties rather than the alleged ones. *Illusions do not lie.*

• • •

In many cases, one believes that the way something looks is an indicator of underlying truths, when it is not. The mansion looks more expensive than it is. The actor looks younger than he is. The boulder looks heavier than it is. In all these cases, one perceives something and one believes one ought to perceive something else. The truth is that, for various reasons, one believes the mansion ought to look less expensive, the actor older, and the stone lighter. Yet, there is no real separation between appearance and reality. One has two different sources of beliefs and thus one can disentangle one's experience—say, hair color—from one's beliefs—say, the actor's age. In such cases, one is aware that the two properties are not identical. Gray hair is likely a sign of age, but it is not necessarily so. People dye their hair or, vice versa, people may be gray-haired and young. However, very often, the property one is more interested in (the alleged property) is not easily accessible. One can access only proxy properties. The alleged property is more elusive.

The alleged property—while more elusive—is not necessarily impossible to perceive, but it may be beyond the grasp of our perceptual skills. Ideally, one moves from cases such as gray hair vs. age and then proceeds to cases such as apparent temperature, mirages, perceptual constancy, metameres, visual illusions, and Benham's top.

Apply now the new explanatory strategy to the case of white hair. Ivan's hair is white. He is twenty. However, if one had the belief that an indicator of old age is white hair, one would think that Ivan looks old, when in fact he is not. As it happens, when one has white-hair, one usually is old too. While nobody can perceive age as such, we can see hair color. Thus, one derives the (wrong) belief that old age mandates white hair. So, when I see someone young and with white hair, I describe what happens in term of illusions, while, in fact, it is only a case of wrong beliefs. I introduce a fictitious illusory appearance to justify the mismatch between my beliefs about gray hair and its presence on a young man.

Because of these limitations, our perceptual system cannot perceive everything we would like. For instance, we cannot perceive the internal composition of matter, or absolute size, or absolute movement, or absolute shape, or the exact angle between two lines, or the age of an individual, or the presence of water. Therefore, our perceptual system settles for second best. Our sensory organs single out *proxy* physical properties that co-occur with desired *alleged* physical properties. Sometimes, things go wrong and one is faced with the difference between what one perceives and what one believes one ought to perceive. In these cases, mistakenly, the classic notion of illusion is called to rescue. Yet, what one perceives is what is there.

Illusions do not differ from standard perception. Illusions allow us to pinpoint the proxy properties we perceive every day rather than the alleged properties we erroneously believe we perceive. In cases of so-called illusions, one perceives a physical property, which is there. However, *illusions highlight the difference between what we actually perceive and what we believe we perceive.* Illusions show that what one perceives during standard perception is not what one believes one ought to perceive.

Consider the case of a mirage (Figure 10). Sometimes mirages have been considered illusions because one sees water where there is no water. One walks in the desert and believes to see a pool of water that will eventually reveal itself as heated sand. In fact, a mirage is not an illusion in a strict philosophical sense but rather a case of standard perception inducing wrong beliefs. One can even take a snapshot of a mirage with a camera. The explanation in terms of illusions derives from the belief that one sees water when, in fact, one does not see water as such. Rather, one sees *proxy* physical properties that co-occur with water. In this case, the capability of reflecting light rays (as would a horizontal mirror) is the proxy property. This property is shared both by water and by super-heated layers of still air—let alone horizontal mirrors. In the wild, such a property is usually instantiated by pools of water that constitute a horizontal partially reflecting surface. Occasionally, this property is exhibited by other materials—e.g., a hot layer of still air. Unsurprisingly, our perceptual system picks up the proxy property, which is easily available by visual means.

Thus, a proxy property is a substitute for what we either hope or expect to find. In the desert, for instance, our target is water. In this case, the alleged property is "being made of water." Such a property is unavailable by visual means. A human body can single out other properties though—e.g., behaving as a horizontal mirror. As a result, in our environment, human beings hold the belief that water looks like a horizontal mirror. Keep in mind, crucially, that both properties—being made of water and behaving as a horizontal mirror—are physical properties. None of them is a phenomenal character.

If you go to the desert, unusual circumstances will take place if compared with a standard human environment. If things were different, of course, different conditions would apply. A pool of water would look like heated sand if one were a Fremen from the fictional Arrakis world as imagined by writer Frank Herbert in his unforgettable saga *Dune*. On our earth, the most common case of horizontal mirroring is offered by pools of water. The heated layer of still air hovering above the sand instantiates the proxy property, the mirror behavior. In this case, the presence of water is not visually perceivable, one is fooled into seeing a horizontal mirror where there is no water. Wrongly, one believes that only water looks like a horizontal mirror. From a practical perspective, this is often the case; the average person does not hunt for water in the desert. Philosophically, the ensuing deflationary stance is wiser. One holds a wrong belief based on what one perceives every day. The mirage does not tell us anything special about what one perceives in the desert. However, it reveals that what one has always believed water looks like, is not unique to water. It is the proxy property that is horizontal mirroring.

A couple of explanatory caveats. First, the relation between a horizontal mirroring and being made of water is contingent: there is no need for any normative causal connection. It would be useful, but it is not necessary. Contingent co-occurrence is enough. Second, as stressed above, the reflection of light-rays is not an intermediate step in the perceptual causal chain between the remote object and the subject, nor a look, nor an appearance, nor a sign, nor an index. It is just yet another physical property.

Thus, the mirage reveals that, whenever one believes one is seeing water, one is seeing the proxy property of horizontal mirroring. No illusory mental property is required by this account. Both the alleged and the proxy properties are just physical properties. Largely, cases of illusion are characterized by the following circumstances:

- *contingently*, one's perceptual system strives to perceive the *alleged* physical property
- *unfortunately*, accessing the alleged property is either impossible or very difficult

- *luckily*, another physical property—the *proxy* property—co-occurs frequently
- *easily*, one's perceptual system perceives the proxy property
- *mistakenly*, one believes one perceives the alleged property
- *unusually*, one perceives the proxy property without the alleged property
- *stubbornly*, one sticks to the belief that the two properties are the same. Because one knows that there is no alleged property, the notion of illusion is introduced to explain why one perceives a property that is not there.

Illusions are a cultural creation rather than a perceptual phenomenon. They are a form of epistemic dictatorship stemming from imposing an arbitrary notion of reality—what things ought to be—and downgrading the world we live in as illusory experience. When the world does not conform to a model imposed as the norm, it is exiled to a subjective domain.

Illusions are not a perceptual phenomenon but delusions. They reveal a tension between what we perceive and what the community establishes we ought to perceive. *Illusions fill the gap between beliefs and experience, between authority and individuals. Surprisingly, beliefs win against actual experience. People set aside their own experience to follow the epistemic authority of the academic community. The theory of spread mind endorses experience rather than historical beliefs about what one ought to perceive.*

So far, I have sketched a draft for how illusions can be modelled in terms of perception using proxy physical properties and alleged physical properties. To convince the reader that this framework is successful, let us review a selection of the most famous cases. This gallery assembles a vast bestiary of perceptual cases to illustrate how the consistent application of the theory of spread mind permits us to locate, in all cases, an actual physical property.

. . .

The notion of apparent temperature is a straightforward example. Nowadays most weather forecast systems inform us not only of tomorrow's expected temperature but also of tomorrow's expected "feels like" or "apparent" temperature. The "feels like" temperature is the alleged temperature one expects to feel, which is different from the temperature as measured by scientific protocols.

The traditional explanatory strategy is always the same: an environmental property chosen by our belief system is taken to be the real temperature while another property—which is equally physical and environmental—is neglected. The former is the alleged property while the latter is the proxy property. Because of the authority of physics, the former is considered to be the real temperature and the latter is downgraded to an apparent or subjective temperature. In fact, historically, a notion of temperature has been chosen—usually, the average speed of molecules or something close to it. Yet, our body is not sensitive to that property. We perceive something else, a chemical reaction in our skin that is often correlated with the average speed of molecules in the environment. The mismatch between the historical notion of temperature and the actual physical phenomenon one experiences endorses the notion of illusory appearance. We are taught what temperature ought to be, and meekly believe it.

Painstakingly, meteorologists have gathered the physical properties that are the real cause of human experience and called it *apparent temperature*. Apparent temperature is not exactly the temperature one experiences, but the set of physical causes closer to what one experiences. Apparent temperature is also a real physical phenomenon, only it is not the one chosen by Fahrenheit and Rømer: it approximates the *proxy property* one perceives, while the temperature of Fahrenheit and Rømer is the *alleged property* one believes one ought to perceive but nobody has ever perceived.

The historical objective temperature is a physically meaningful and practically useful notion, only nobody experiences it. What one experiences is a more articulate physical property that, because of ontogenetic and phylogenetic factors, produces effects by means of human bodies. Consider an approved definition of apparent temperature. The National Digital Forecast Database (NDFD) states that for temperatures greater than 80°F,

Apparent Temperature: The perceived temperature in degrees Fahrenheit derived from either a combination of temperature and wind (Wind Chill) or temperature and humidity (Heat Index) for the indicated hour. When the temperature at a particular grid point falls to 50°F or less, wind chill will be used for that point for the Apparent Temperature.

The structure of this definition reveals to what extent meteorologists have struggled to put together a combination of various phenomena such as the standard temperature and the humidity. Nevertheless, the definition has only a limited accuracy because different human bodies react differently to the same external physical properties. Thus, different bodies allow slightly different physical properties to take place. The case of apparent temperature shows clearly that a physical property—causally singled out by human bodies—is actually perceived, and another property—promoted by science, common sense, and other historical factors—is taken to be the real one. Of course, the latter property is neither more physical nor more objective than the former. Both apparent temperature and Fahrenheit's temperature are physical phenomena.

·   ·   ·

Consider what is likely the most famous case of visual illusion, namely the Müller-Lyer illusion, which is usually drawn as two double-headed and stylized arrow-like figures (Figure 10). The upper line segment— the shaft of an arrow with two pointed heads—appears shorter than the lower line segment, the shaft of an arrow with two finned tails. However, both lines have identical physical length, as can be ascertained with a ruler. The standard view suggests that what we see is an illusion which consists in seeing the upper segment as shorter than the lower. Thus the length we see is deemed to be illusory and mental.

The solution I propose is the same that we have seen in all other cases. An alleged property, namely the absolute length of a segment, is masked by a more complex but easier to pick up proxy property. Our visual system uses the latter to estimate the former. We perceive the

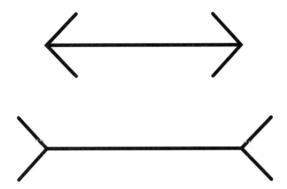

Figure 10. The Müller-Lyer illusion. Isn't the lower line really longer?

proxy property but we believe we perceive absolute length. The notion of illusion comes to the rescue to save our beliefs.

The proposed solution will become obvious once one considers the nature of the perceptual processes that allow us to grasp the dimensions of objects. As is well known, vision is hampered by countless issues such as unknown optical parameters, retinal non-uniform structures, unknown relative ego-position, and so forth.[48] Vision is a well-known mathematical ill-posed problem that does not admit exact solutions. Visual systems have no physical means to measure the absolute length of external objects: human eyes do not project laser beams capable of measuring absolute length. However, natural selection always looks for workarounds. In this case, the workaround consists in looking for an alternative phenomenon co-occurring. Unfortunately, in the case of size, there is not a single phenomenon and so our visual system resorts to collecting a long list of cues—e.g., relative size on the retina projection, homogeneous shapes, textures, patterns, concave and convex angles, presence of parallel and converging lines, and so forth. These properties *often* co-occur with lengths and sizes of external objects. In a standard human environment, most angles happen to be right angles and straight lines happen to be either parallel or to cross at right angles. Similarly, objects with a similar shape often have the same size. Moreover, it turns out that lines on convex objects appear shorter than lines on concave objects, as in the case of Müller-Lyer-like arrowheads. And so forth. The list is long. Although such a list varies, in human beings it shows a certain uniformity because they share the same environment and the same body structure.

Bottom line: when we look at the Müller-Lyer segment, we see two lines of different lengths. It happens because the two lines differ in the property our visual system uses to estimate length. The two lines are not different in their absolute lengths. They differ in respect to another physical property, though, which is the one our visual system picks up to estimate size and length. So, the upper line is really longer than the lower one in the sense that what the visual system perceives is a proxy length and not absolute length; it is *proxy longer*. We perceive another property, which happens to be greater in the lower segment than in the upper one.

. . .

Another interesting case is the Enigma illusion (Figure 11, left), devised by Isia Leviant in 1981, in which a static pattern is perceived as though it is moving. Several static patterns have this property. Drift illusion is one famous example first exploited in 1999 by two perception scientists, Jocelyne Faubert and Andrew M. Simon. The same structure has been later exploited in many popular pictures.[49] In short, the standard account is that one has an illusory perception of motion, while no motion occurs. Allegedly, one sees illusory mental motion. As usual, it is supposed that motion, since it is not instantiated by the object, is instantiated either in the mind or in the brain or both.

However, both in the Enigma illusion and in analogous cases of peripheral drift illusion, certain static patterns cause the activation of the same neural areas as physical movement. Both actual motion—displacement of objects in time—and other equally physical phenomena are among the causes that trigger a certain response in one's brain.[50] Among such physical states of affairs, Leviant singled out one of those rare static phenomena that share the same property that our visual system picks up in the case of moving objects. The explanation, once again, is that a common physical cause triggers the same physical response. If two keys, allegedly different, open the same lock, do they not share the same causal property?

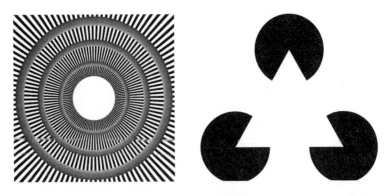

Figure 11. Enigma illusion (left), Kanizsa triangle (right).

The human visual system does not detect motion per se. Our visual system does not project laser beams that keep track of the actual movements of objects. Our visual system tries to take advantage of certain visual properties. It singles out properties that reliably co-occur with movement. Some of these properties occur also in static patterns. Therefore, in the case of the Enigma illusion, the alleged property is actual rotary motion while the proxy property is the complex set of gradients. Such a set of gradients is a common outcome of moving objects. In normal circumstances, this distribution of gradients does not occur in static patterns. As a result, when the system picks up such a proxy physical property, the system will pick up rotary motion too. Sometimes, though, a static pattern exhibits these kind of gradients—e.g., the Enigma illusion.

As in the case of temperature, physical motion is the alleged property imposed by the authority of science and common sense. A complex set of static gradients, geometric relations, and spatial derivatives is the proxy property our visual system singles out.

. . .

In the 1950s, Italian psychologist Gaetano Kanizsa devised many ingenious figures that dramatized the phenomenon of illusory contours. Among them, the Kanizsa triangle became the most popular (Figure 11, right). It is neither a hallucination nor something that appears hovering above the pattern. In a sense, one perceives the triangle but one does not see it. However, one perceives a certain *triangleness* in the middle of the figure, so to speak. It is the well-known phenomenon of visual closure or completion.[51] The neural evidence is coherent with its phenomenology—the corresponding cortical activities are akin to those that would result from staring at a standard triangle. As usual, one is tempted to fall prey to the standard account of illusions—namely to assume that while no physical triangle is obtained, one's brain concocts an illusory mental triangle. The alleged property is the ideal triangle, a triangle no one has ever seen.

Once again, though, the theory of spread mind sheds a new light. In practice, since the visual system has no way to perceive complete objects or shapes, it settles for second best. The proxy properties that the visual system picks up to test whether a triangle occurs, which is the alleged physical property, are identical with a huge set of visual cues. Thus, the proxy property picked up by a standard human visual system is something like the existence of three visually salient angles whose sides approximatively conjoin, and not the objective alleged property which is something like a plane figure with three straight sides and three angles. The proxy property, which is a physical property, is shared by many more physical phenomena than Euclidean triangles alone. In this case, the three blobs instantiate the proxy property but not the alleged one. Kanizsa's three blobs are a key that opens the triangle lock, so to speak.

The bottom line is that the Kanizsa triangle is among the phenomena we pick up when we perceive triangles. However, it does not conform to the ideal geometric notion of triangles. In everyday perception, when we see actual triangles, what we see is the proxy property and not the alleged property. The proxy property is good enough for everyday life. Of course, when one is confronted with a Kanizsa triangle, the situation is unusual enough that the alleged property does not occur.

. . .

Colors are another nice example of contrast between beliefs about what we expect we ought to perceive, and what our perceptual system actually singles out. This is not to say that a color is created by one's perceptual system, only to state that one's perceptual system carves out a certain set of physical phenomena. Such a set fixes what we actually perceive. The color we perceive is physical, but the criteria for its selection are arbitrarily fixed by the causal structure embedded by our visual system and body.

Color illusions arise from the conflict between two equally powerful forces: on the one hand, a very authoritative cultural and scientific series of models of colors,[52] and on the other, an extraordinarily complex biological perceptual system. As a result, laypeople, scientists, and philosophers are faced with a very persuasive model of what colors are expected to be. At the same time, though, the attempt to understand what the human color system picks up has so far defied most neuroscientists and psychologists.

All color illusions can be addressed using the explanatory strategy deployed above. One is confronted with a situation in which one sees a color, say, red, and one is taught that that object is not red but another color. On what basis is one taught so? Because another authority—likely science—supersedes experience with a belief about what one ought to see.

Once upon a time, colors were simply anything that looked colored: white and black as well as brown, gold, silver, pearly, nacre, brassy, imperial purple, and so forth. As of Newton's model of colors, every student has been taught that colors are light frequencies.[53] As a result, whenever these notions do not match one's perception, illusions are invoked to explain the mismatch between Newton's authority and experience! Thus, if we look at a pattern that does not contain any frequency in the range between 610 and 700 nm and nevertheless we see red, we will be taught that we are victims of an illusion because—so authority tells us—no *real* red is there. Yet, we perceive a physical property that is shared both by the light rays of frequency 680 nm and by other combinations of patterns that do not contain that frequency. If it were not the same physical property, it could not produce the same effect in us. Therefore they must share the same physical property. Only one class of phenomena qualifies to be, say, red: namely, those that are red when we look at them. I stress the point—only those phenomena that *are* red when we look at them, not those that look *red* to us.

. . .

Metamerism occurs when perception borders on illusion. Allegedly, color metamerism occurs when subjective colors match different spectral power distributions. Thus, metameres are akin to illusions insofar as one perceives something that does not match the external world. The term metamere is used for color properties but the same notion can be applied to other sensory modalities. For instance, in the food and cosmetic industry, flavoring substances are chemically different molecules that produce the same olfactory or gustatory experience. The widespread wisdom is that metameres are *different* in physical terms but *appear* the same in one's experience.

A simpler account is available: what one perceives is not the alleged true color or true taste. *Every time one's experience is the same, one experiences the same physical property.* Metameres are identical physical stimuli that, not surprisingly, cause the same combination of photoreceptor activations. For historical and cultural reasons, though, such physical stimuli are confused with other properties instantiated by different physical phenomena. As a result, one believes that different stimuli produce identical experiences, while this is not the case. The stimuli, too, are the same.

First, consider color. For the sake of the argument, suppose I adopt a simplistic color model that dictates that colors correspond light power spectra. Call it the dominant frequency. In such a belief system, colors ought to vary whenever such a spectrum varies. Yet, we know that many color spectra with different dominant frequencies are perceived identically. The mystery is easily solved, the dominant frequency is the alleged property, and the cause of our identical perception is the proxy property. In real cases, color models never completely match the subtleties of human color perception. We settle for reasonable approximations ranging from dominant frequency up to more complex models. The notion of color metamerism is based on the notion of an alleged true color while such a notion is only hypothetical.

Consider taste. When I eat strawberries, I experience strawberry-taste. I assume that strawberries must have a property that triggers my experience. I know only that strawberries are such that I experience strawberry-taste whenever I eat them. Eventually, because of some historical development, a science of taste develops. The savants of this discipline teach me that tastes are chemical substances and I believe it. They teach me that, *in reality*, the taste of strawberries is a certain molecule. I believe that too. Now, I find another molecule, much less expensive than real strawberries. If I eat the cheaper molecule, I will experience the same taste I do when I eat strawberries. Strawberries and the new substance are gustatory-metameres since they are different but "appear the same." My brain gives two different substances the same interpretation. Right? Not at all! The traditional explanation is misleading. It ought to be obvious that, if strawberries and the new molecule produce the same effect in one's perceptual system—which is some kind of physical structure and not an immaterial soul—they must share a common physical feature. Again, *if I have two keys that look different but open the same lock, they will share a physical feature that is the cause of the lock opening.* This common physical feature is shared both by the strawberries and by the new cheaper substance. It constitutes my experience. The common feature is the property that my gustatory system picks up.

In the case of tastes, the alleged physical property is some theoretically devised property, while the proxy property is something shared by both strawberries and the new substance. In the case of colors, the alleged physical property is, say, the light frequency, and the proxy property is a rather complex set of features owned by a patch color spectrum and its surrounding areas. Historically, it is informative to note how often, due to practical factors, a property has been confused with another one. By and large, all illusions could be cast in terms of metamerism. A pool of water and a heated layer of sand are visually metameric.

• • •

Perceptual constancy is akin to metamerism. Once again, we have beliefs about what one expects to perceive and what one actually perceives. For instance, I look at a sheet of paper reflecting yellow light and I perceive it as though it were (almost) white. I look at Emily from a great distance and I perceive her as being as tall as myself. I observe a coin that I turn between my finger and I perceive it as being perfectly circular no matter its orientation. Very succinctly, the notion of perceptual constancy is based on that of proximal stimulus. The proximal stimulus changes while the percept and the external object are constant. In fact, in the case of perceptual constancy, the variations are not in the properties of the physical object, but in the property of the effect the physical object exerts on one's sensors, something akin to the vintage notion of proximal stimulus.

Consider the white sheet, under a yellow light and then again under a white light. According to the tradition, different physical phenomena produce an identical experience thanks to clever internal computations. Such a conclusion is not necessary. Alternatively, if two situations produce the same effect inside one's body, they must have something in common. The common factor is the actual cause of one's experiences. Under different lighting conditions, the same sheet of paper produces the same effect inside one's body. The notion of perceptual constancy is required only if one assumes that the cause of one's experience is the proximal stimulus rather than the actual object.

A caveat. The effect I speak of is not a phenomenal experience but a physical effect caused by the sheet of paper. For instance, in both cases, one utters "pure white" or one points to the same swatch in a Pantone palette. If one can do that, inside one's nervous system, at a certain point, the same effect has been elicited by the sheet of paper under different light conditions. Therefore, the two physical circumstances, as regards their causal efficacy, are equivalent. Thus, they share a common property that is the actual cause of one's experience. Such a common property is the proxy property.

In sum, perceptual constancy—which is of enormous interest from a cognitive or psychological perspective—is not an issue here. The conceptions of metameres and perceptual constancy are cultural artifacts rather than perceptual phenomena. They both contrast what one perceives with what one *ought to* perceive because of historical models of perception.

*   *   *

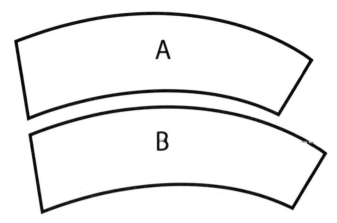

Figure 12. Jastrow illusion.

The list of illusions could fill an entire book. I hope the general structure of the explanatory strategy is clear enough: *illusions are misbeliefs about what we perceive and about what we believe we ought to perceive.* Such a strategy can be exploited with many more beautiful illusions that amaze and amuse. Each of these cases will tell us something interesting about the difference between what one actually perceives and what one believes to perceive.

So far, I have focused mostly on cases in which an illusion is the result of the presence of a proxy property. Sometimes, though, illusions arise from the absence of the proxy property. It is instructive to consider at least one case of this kind. Consider the Jastrow Illusion (Figure 12), discovered by Joseph Jastrow in 1889. Two shapes are identical but look different. Beholders see the figure A as though it were smaller than B and thus of a different form since the original form could not be maintained by such a change in size. However, the illusion is obtained by absence rather than by presence of the proxy property. The pair of figures lacks the proxy physical property, which is the target of standard perception in other situations in which the external figures instantiate the alleged physical property—namely, having the same size.

Every illusion can be explained adopting the same explanatory strategy: i.e., finding the neglected proxy property. The proxy property is the real target of perception and, in *unusual* circumstances, does not occur together with the alleged property.

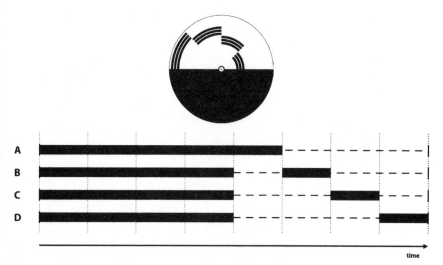

Figure 13. Benham's top. If the disk is spun clockwise the colors are red, yellow, green, blue; counterclockwise, the colors are A=blue, B=green, C=yellow, D=red (from the innermost outward).

•　　•　　•

In 1894, English toymaker Charles Benham sold a circular top painted with black and white patterns that, when the disk is spun, are perceived as a series of arcs of pale color. The top realizes a remarkable color illusion. Apparently, its achromatic surface lacks any color and yet, when it is spun, it appears as though it were colored (Figure 13, top). It is undeniable that, when it is spun, one sees various colors. For the sake of the example, consider the pale red of the inner circle. How is it possible? Where is the pale red one perceives? Since the top is achromatic, is it not a mental color? Surprisingly, I believe it is possible to prove that one sees red because, when the disk is spinning, the disk is indeed pale red.

According to tradition, Benham's top is colorless but, when it is spun, it appears as though it were colored. The traditional account is that one perceives a property that is not physically there. No pale red occurs on the disk surface. The pale red is only in one's illusory experience (whatever that means). One perceives a mental red. Allegedly, and suspiciously all too conveniently, such a red is taken to be generated in the brain.

According to the theory of spread mind, Benham's top is indeed colorless when it is still. However, when it spins, it is pale red. How is that possible? The solution is to revise our notion of pale red. Pale red is instantiated not only by pale red static patches, but also by the spinning disk. The property in common between a pale red patch and the spinning disk is the proxy property. The spinning disk, thus, is pale red. The spinning disk is physically different from the still one and instantiates different physical properties.

In comparison with other known illusions, Benham's top is tricky for a remarkable fact. It exploits a common misconception, namely that the real properties of an object remain the same regardless of whether the object is moving or not. In other words, it is assumed that the physical properties of the still disk must be the same as the physical properties of the spinning one. Yet, such an assumption is obviously false. A moving object and a still one have different properties.

For instance, if I heat up an object, I observe that, as long as its temperature increases, its color changes too. Now, heat is average molecular speed. Why should I be surprised that a moving (i.e., spinning) disk had a different color than when it is still? Consider the iridescent wings of certain butterflies, in which hue changes in proportion to the angles of observation and illumination. Iridescence is caused by multiple reflections from two or more semi-transparent surfaces in which phase shift and interference of the reflections modulate the incidental light by amplifying or attenuating some frequencies more than others.[54] In such a case, the color difference is the result of a change in the orientation of the microstructures in the wing. In the case of Benham's top, the color difference is the result of a change in relative momentum. Physical changes trigger different physical properties. Bottom line, there is no real reason why an object should not be pale red when spinning and achromatic when still. A change in physical properties results in a change in other physical properties.

But why does spinning Benham's top instantiate colors? We need to dig more into the details of its mechanism. A promising line of enquiry considers the role that temporal integration and movement play. Consider for a moment a less demanding case—namely, being gray. Take a simple spinning disk made of alternate white and black rays. There is nothing gray when it is still. However, when the disk is spinning at a speed that is higher that the temporal integration time of one's photoreceptors, one sees a gray disk. Is there a gray disk? Yes. The property we call "gray" is instantiated by the spinning disk but not by the still disk. If one sets aside parochial beliefs, one no longer resists the fact that a spinning disk is gray and, when still, the same disk is black and white. A spinning black and white disk is gray relatively to a visual system with a slow integration time. Gray is a physical relative property, like all the rest.

One might object that the property of being gray depends on the integration time of the beholder's visual system. Fair enough. Nevertheless, this is comparable to the white and black rays on the disk surface that are white and black only in relation to various parameters of the beholder's visual system. For instance, if the beholder had many more kinds of cones in the retina, it is likely that what one perceives as white would no longer be white. If white is taken to be a case in which all primary components have identical intensity, it is unlikely that, if I could sample the spectrum with different primary components, I would obtain identical values. On the other hand, if one had a much higher spatial resolution, any surface would dissolve into gazillions of multicolored particles. In short, the black and white still disk is no less dependent on the oddities of my visual system than the gray disk that I see when it spins. Moreover, the disk is as much in a natural state when still as when it is spinning. Therefore, the objection is refuted: the gray does not depend on the beholder's temporal integration time, rather the gray is perceived *thanks to* the beholder's temporal integration time. All physical properties depend on the right causal circumstances for their occurrence. They exist relatively to other physical systems.

Do a mirage and Benham's top fit into the same category? They are somewhat different. Supposedly I can take a picture of a mirage but I cannot take a picture of the pale red of the spinning top. Surprisingly, this is not true. In fact, if I use a slow enough aperture time and I take a picture of a black and white spinning disk, my snapshot will show a gray disk. Furthermore, I can envisage a camera mimicking some of the shortcomings of the human retina. If my camera had different integration times for different color components, it would shoot colored photographs of spinning Benham's disks.

In fact, a phenomenon akin to Benham's top afflicts most commercial digital cameras—namely, the Moiré effect. It is a consequence of the Bayes pattern used to interpolate colors out of adjacent pixels, and produces undesired "illusory" color effects from achromatic patterns. Yet, the Moiré effect does not require illusory colors inside cameras. It shows that commercial cameras do not pick up colors, but rather composite properties out of adjacent points. Cameras, too, have their own proxy and alleged properties. Therefore, sometimes it is possible to take a snapshot of a "pale red" that apparently is not there, but, given the right conditions, takes place. The composite properties are part of the cluster of properties that the camera picks up.

In the case of mirages, one can easily access both objects: on one side, the water and the reflection; on the other side, the sand and the reflection. As a result, they seem relatively independent. In the case of Benham's top, the proxy physical property is so intertwined with the oddities of one's perceptual system that it is very difficult to single it out. When Benham's top is spun, one is not confronted with two easily separable properties, but rather with a manifest proxy property—the pale red—and with a conceptually clear but disguised alleged property—the black and white pattern. Because one is unable to conceptualize the relation between such properties, one is tempted to explain them by means of illusory perception. The bottom line is that traditional explanations based on mental appearances are bad explanations. They encourage laziness.

5.

Hallucinations and Dreams

> The inventions of alternatives to the view at
> the center of discussion constitutes an
> essential part of the empirical method.
>
> —**PAUL FEYERABEND, 1975**

N THE PREVIOUS chapter, we tackled illusions and we saw that they are perceptions of actual physical properties. We can now tackle the other traditional obstacles any strong realist view of experience faces— namely, hallucinations and dreams. In fact, if the traditional account of hallucinations were sound, hallucinations would prevent any strong identity between experience and world, between mind and world, and between appearance and reality. Famously, in his monumental 1890 work *The Principles of Psychology*, William James stated (italics mine):

> A hallucination is a strictly sensational form of consciousness, as good and true a sensation as if there were a real object there. *The object happens to be not there, that is all.*

If one could have an experience of something that does not happen to exist at all—as James states in the above passage—the presented theory would fail. Since James's time, most scholars have assumed that hallucinations are akin to perception but for the absence of external objects. Having an experience without an external object is, if true, the main reason to distinguish mind and world. If hallucinations occurred when no objects happen to be there, experience would stem out of an inner mental domain. Similar formulations have been put forward by most philosophers, psychologists, and neuroscientists—they all stress the absence of the external object.[55] In this regard, Oliver Sacks wrote in his 1992 book about hallucinations:

> Precise definitions of the word "hallucination" still vary considerably, chiefly because it is not always easy to discern where the boundary lies between hallucination, misperception, and illusion. But generally, hallucinations are defined as percepts arising in the absence of any external reality—seeing or hearing *things that are not there.*

Yet, this widespread belief is only allegedly supported by empirical evidence. Surprisingly, *when one hallucinates, one sees and hears things that are there.* We only need to revise the simplistic notion of *there.*

The traditional account of hallucination is neither an *a priori* truth nor a conceptual possibility. It is an empirical fact that, as such, must be based on empirical evidence. Such evidence, I will argue, does not exist. If the empirical evidence has been overestimated, the whole appearance vs. reality house of cards will tumble down.

Can hallucinations, like illusions, be cases of actual perception? Is there a physical object whenever one hallucinates? Here, I show what and where such an object is. If hallucinations, too, are perceptions of physical objects, the traditional account of perception as reliable hallucination can be discarded. On the contrary, *I will turn upside down the traditional direction of explanation: perceptions are not reliable hallucinations, rather hallucinations are spatiotemporally reshuffled perceptions.*

The key to succeed is to exploit the causal geometry of experience we have outlined earlier. Events are spread and are causally carved out. The world we live in and our experience are the result of such a carving, world and experience being the same. Perceptions and hallucinations are different carvings of the world. Both perception and hallucination, though, are fragments of the world. The theory of spread mind shows how perception spreads to hallucinations. Hallucinations, being forged by the same causal forces that single out perception, are not inhabitants of an inner arbitrary world. Hallucinations do not stem gratuitously from the immaterial space of an inner mental enclave. Hallucinations are cut out from the same stuff our experience is made of: namely, the physical world. The way in which they are cut is less ordered than everyday perception but the way they are cut is not metaphysically different.

• • •

Hallucinations are not all the same. This statement is not so obvious from a philosophical viewpoint. In fact, philosophers as diverse as William James, Bertrand Russell, and A. J. Ayer have often adopted a simplified notion of hallucination that amounts to the notion that hallucinations are phenomenal experience without an external object. Most authors check whether the thing one perceives is immediately present. If the object is not, the experience is a hallucination. Such a conception is questionable for at least a couple of reasons. First, as we will see, the notion of "there" is vague, imprecise, and parochial. Second, it is too coarse since it does not differentiate among hallucinations. In this section, I address the latter concern.

Hallucinations are not as bizarre as they are often assumed to be. Additionally, they are not all bizarre to the same extent. Some hallucinations are extremely close to everyday perception. Others are apparently more exotic. Between perception and the most bizarre hallucinations, there is a continuum of cases. Hallucinations and the everyday world differ by degrees.

For instance, if I hallucinated seeing the chipmunk I came across a few months ago in Boston, it would be less amazing than if I had hallucinated seeing an animal I have never observed. By the same token, it would be even more amazing if I hallucinated seeing a color outside my standard color space and still more amazing if I hallucinated seeing something nomologically impossible. To cut a long story short, different hallucinations have different degrees of separation from everyday life.

A taxonomy of hallucinations based on their distance from the hallucinator's world might be helpful to shed a new light on their nature. Here I organize hallucinations into classes based on their causal proximity with everyday life. Such an effort might be rewarding if it turned out that some of these classes are empty.

The first class is everyday perception. One perceives things as they actually are—e.g., Emily perceives a red apple and a red apple is there. It is the Shangri-La scenario. It is easy to cope with using the experience-object identity I defend here. In the theory of spread mind, the explanation of everyday perception is straightforward—we experience a red apple because our experience is the red apple.

Second is the case of illusions. One perceives something differently from what it is. More correctly, one believes that one perceives something differently from what it is. For example, I perceive an apple as red but have reason to believe the apple is green. This case too, as we have seen, can be reduced to actual physical properties that exist and that happen not to match one's beliefs. Illusions can be solved by revising one's beliefs about the properties and the objects that one perceives.

Third and finally, we face the huge class of hallucinations. I propose to divide them into *ordinary* and *extraordinary hallucinations*. Ordinary hallucinations are experiences of objects or of parts of objects that one has experienced before. Extraordinary hallucinations are experiences of objects or parts of objects whose components one has never experienced before.

My key claim is that extraordinary hallucinations do not exist. They have never occurred. They are only a scientific myth flanked by philosophical views. Conceivability is not a sufficient proof. Modal logic will not help us here. We will need to address the available empirical evidence. We need to know if people have ever had a hallucination whose elementary components were not caused by the external world, and thus that was native of a pure mental world. If no empirical evidence supports the existence of extraordinary hallucinations, the argument from hallucinations is doomed. In fact, ordinary hallucinations can be explained in terms of reshuffled perception.

Ordinary hallucinations can be further divided into two subgroups. The first group comprehends cases in which one experiences things that do not exist when one experiences them but that one has perceived *elsewhere* and *at some other time*—e.g., Emily experiences a red apple and no red apple is there, but she has seen red apples before. The second group comprehends cases in which one experiences things that do not exist when one experiences them but whose properties one has perceived *elsewhere* and *at some other time* albeit in a different order—e.g., Emily experiences a red apple and no red apple is there, but she has seen red patches and green apples before. The crucial feature of such hallucinations is that they are the result of the spatiotemporal reshuffling of actual objects. Thus, ordinary hallucinations can be explained in terms of objects. They do not require any mental world.

In contrast, *extraordinary hallucinations* address cases in which what one experiences cannot be reduced to any previous experience. They, too, can be divided into three subgroups. The first group comprehends cases in which one experiences things that do not exist when one experiences them and that one has never experienced but that one might have experienced—e.g., Emily experiences a red apple and no red apple is there. Moreover, she has never seen red objects or apples. Hume's missing shade of blue belongs to such a subgroup. The second group comprehends cases in which one experiences things that do not exist when we experience them, that we have never experienced, and that we could not have experienced—e.g., Emily experiences an infrared light from the apple. Due to her sensory limitations, she cannot possibly perceive infrared. Nonetheless, infrared is a physical phenomenon and, if her body were different, she might perceive it. Finally, the third group is the most extreme—one experiences things that do not exist when one experiences them, that one has never experienced, and that cannot be experienced by means of an actual experience—e.g., Emily experiences a four-sided triangle.

Ordinary hallucinations are nothing but delayed reshuffled perceptions, while extraordinary hallucinations simply do not exist. To show that such is the case, ordinary hallucinations are considered separately from other classes of hallucinations. Ordinary hallucinations are constrained by one's actual experience. Only extraordinary hallucinations seem to depart from the physical world. Luckily, they do not exist. Thus, the claim is twofold:

- ▪ Ordinary hallucinations are a form of perception
- ▪ Extraordinary hallucinations do not exist

Standard perception	Experiencing things that are there.	I experience a red apple and the red apple is there.
Illusion	Experiencing things that are there, but differently.	I experience a red apple and a red apple is there but I have reasons to believe it is green.
Ordinary hallucination I	Experiencing things that do not exist when we experience them but that we have perceived *elsewhere* and *at some other time*.	I experience a red apple and no red apple is there in front of me, but I have seen red apples before.
Ordinary hallucination II	Experiencing things that do not exist when we experience them but whose properties we have perceived *elsewhere* and *at some other time* albeit in a different order.	I experience a red apple and no red apple is there and I have never seen one, but I have seen red patches and green apples before.
Extraordinary hallucination I	Experiencing things that do not exist when we experience them and that we have never experienced but that we might have experienced.	I experience a red apple and there is no red apple. Nevertheless, I have seen neither red objects nor apples. This is, of course, Hume's missing shade of blue.
Extraordinary hallucination II	Experiencing things that do not exist when we experience them, that we have never experienced and that we could not have experienced.	I experience an infrared light from the apple. Due to my sensory limitations, I cannot possibly perceive infrared. Nonetheless, infrared is a physical phenomenon and, if my body were different, I might perceive it.
Extraordinary hallucination III	Experiencing things that do not exist when we experience them, that we have never experienced and that cannot be experienced by means of an actual experience.	I experience a four-sided triangle.

Table 1. Ordinary and extraordinary hallucinations.

The first empirical prediction that these claims put forward is that, during hallucinations and their cognates, one experiences only objects and properties with which one has had physical contact during one's life.

If extraordinary hallucinations were real, they would be metaphysically more demanding than ordinary ones. In fact, they would not originate from the physical world. They would entail that what one experiences is independent of the physical world, possibly even impossible within the physical world, either nomologically or contingently. Only extraordinary hallucinations, if they were real, would provide support for the argument from hallucination. On the contrary, if it turned out that hallucinations are always ordinary, they could be explained by stretching perception so to encompass a larger-than-usual spatiotemporal span. The causal geometry of experience outlined in the previous chapter has the resources to cope with ordinary hallucinations.

· · ·

The first step is to show that ordinary hallucinations are nothing but perception of spatiotemporally composite objects. If one hallucinates objects that one has met before, the distinction between hallucination and perception blurs. A continuum of cases from everyday perception to hallucinations, dreams, and memory can be envisaged. The distinction between hallucination and perception is akin to the distinction between a hill and a mountain. The difference is of practical importance but of no metaphysical relevance.

The core idea is that, if perception of an object is identity with an object that is the cause of current brain activity, hallucinations are perception of objects that occurred in the past and that are still the cause of brain activity. Such a past is still part of one's present. After all, the alleged present is never synchronous with neural activity.

Consider standard perception. I perceive a face. Such an object is constituted by a collection of spatially distributed features. I perceive Leporello's aria by Mozart. It is constituted by a collection of temporally distributed features—the notes and the sung words. Consider Salvador Dalí's portrait of Voltaire in which different entities—two nuns, an arch, and a few Arab boys—are so arranged as to compose Voltaire's portrait. I perceive Ursa Major. Its seven stars are scattered in space and time. In all these cases, everyday objects spread in space and in time.

I have repeatedly stressed that perception is not instantaneous: it takes space, time, and causal steps to complete. The objects we perceive are not instantaneous. This should not be surprising since causal processes can pick up previous causes wherever and whenever they like. Thus, if I have seen, say, a red apple during my life, it is feasible that, if the causal process is still around, I will again see the red apple sooner or later. I will not see *an image of* the red apple. I will see *the red apple* as I saw it the first time. When this will happen, I will call my experience a dream, a memory, or a hallucination. Yet, it will still be perception.

Turning the causal argument upside down, if perception takes place in time, so does a hallucination. If everyday perception takes place notwithstanding a time gap, time is not a critical factor. Thus, hallucinations of objects and their properties are cases of unusually postponed and reshuffled perceptions. The proposal is that, whenever one hallucinates a red apple, it is either because one has previously seen a red apple or because one has seen red objects and apples. The red apples one hallucinates are thus spatiotemporally composite causes of brain activity, no matter where and when their components got in contact with one's body. The resulting prediction is that *one can hallucinate something only if it is made of objects and properties one has met before.*

Let us delve further into the similarities between perceptions and hallucinations. Consider this example. Bob is a normal perceiver but, due to some unusual disease, his perceptual processes are delayed by means of additional neural processes. At the beginning, the delay is barely noticeable. Unfortunately for Bob, such a delay increases with time. After a while, he has trouble interacting with fast moving objects. However, he can still deal with static and relatively slow moving objects. His world has shrunk to the Umwelt of a snail. Despite the severity of his condition, he still perceives the world, albeit in a delayed manner. Furthermore, as shown by control theory, he can cope with arbitrarily long delays by introducing a correspondingly long delay in his actions. Bob copes with his disease by slowing down his actions. Although social life with his peers is now painfully slow, he still experiences the world. The moral of Bob's predicament is that no amount of delay transmogrifies Bob's perception into hallucination. Similarly, one can envisage a continuum of intermediate cases from everyday perception to dream, memory, and hallucination. Since Bob does not hallucinate, neither do I when I experience, say, the apple I ate yesterday.

Perception is a causal process that needs space and time to complete. The amount of space and time is arbitrary. It does not jeopardize the perceptual nature of one's experience. Certain perceptual processes, due to various physical conditions, need a longer time span to complete.

Average standard perception requires 100-300 milliseconds to complete. Occasionally, processes can spread across longer time spans—minutes, hours, days, months, and years. We call such cases memory, dreams, and hallucinations. Yet, they are just cases of perception.

Why are certain experiences more easily considered perception? The answer is practical in nature. Given the action loop that my behavior is part of, it is customary to consider as perception only those processes that allow me to interact efficaciously with their objects. Thus, it is customary to consider as perception only those processes that are fast enough to allow me *to fight back*, so to speak. It is like shooting at something that is very far away. If it is too far, it is in the past. Yet, there is no magic threshold. From a purely perceptual perspective, perceiving an object 300 seconds or 300 days after the light has bounced on its surface is akin to perceiving it after 300 milliseconds, just as it happens in everyday perception. I challenge anyone to find logical conditions or physical constraints drawing a boundary between physical processes of 300 milliseconds and physical processes of 300 days. Once we admit—as any serious physicalist should—that a causal process embeds the relation between the red apple and my neural activity, nothing distinguishes perception from ordinary hallucinations. Causal processes have no expiration date. If perception occurs given a time span of 300 milliseconds, it occurs also given any span of time, as long as the same causal conditions are satisfied. The causal process embodies a relation between an object and my body. However, my experience is not the final neural effect. My experience is the original external cause. My experience is the object.

What separates standard perception from experiences such as memory, dreams, and hallucinations? A practical difference. When I perceive the red apple on the table, I can grab it and eat it. When I hallucinate the red apple, I cannot eat it. Consider an ordinary hallucination of the first kind. I experience the red apple that was on the table a week ago. Of course, I neither grab it nor eat it. The difference between the two cases is practical. Consider an ordinary hallucination of the second kind. I experience the red patch I perceived, say, one week ago in New York, next to the green apple I perceived, say, a month ago in Rome. The red apple made of the red patch I saw in New York and of the apple-shaped object I saw in Italy is a spatiotemporally composite object. A gerrymandered object is the cause of my perception. The Italian-American apple is as physical as the red apple—they are both causes of my brain's activity. However, I cannot eat the Italian-American apple. Thus, I consider it to be a hallucination. It makes a lot of sense from a practical perspective. The red apple on the table now is spatiotemporally spread too, but the space-time span between it and my body is much smaller. The apple I hallucinate is made of elements spread in space and time. The apple I perceive, too, is made of elements spread in space and time. They happen to be closer to my body. No metaphysical difference separates perception from ordinary hallucinations. Customarily, we call *perception* the experiences whose objects we can manipulate. We call *dream, memory, hallucination,* and *astronomic observation* the experiences whose objects we cannot manipulate. In a sense, hallucinations are perceptions of a limited practical value.

The lack of a substantial difference between the two cases—apart from the practical impossibility to eat or grab an object in cases of ordinary hallucination—explains why no introspective difference distinguishes hallucination from perception. No introspective difference marks one's experience. In both cases, one perceives a red apple. During everyday perception, one perceives the red apple that was on the table 300 milliseconds earlier—which is still there, given "the average speed of apples in my environment." During memory, one perceives the red apple that was on the table, for instance, in November 2014, 300 days earlier. Finally, during hallucination, one perceives the red apple that was in New York mixed with the green apple that was in Rome. Perception, memory, and hallucination are processes similar in kind and yet different in temporal extension.

Finally, it is understandable why action-related perceptual processes enjoy a special reputation and status. After all, if we were constantly bamboozled by the red apples that we met during our life, we would oversee the ones in front of us and, as a result, we would starve and die, hopelessly lost in an eternally perceivable past. Daydreaming has its perils. It is conceivable that all kinds of mechanisms must prevent us from being overwhelmed by temporally distant events. They might disrupt our pressing daily interactions with the environment. For most practical purposes, it is better to live in a short present than in the—more or less—remote past.

To convince yourself that a hallucination is, to all extent, postponed perception, consider the most often quoted case of internally produced hallucination—namely, Wilder Penfield's direct stimulation of the brain.[56] Penfield's empirical evidence has been crucial to shaping the notion that hallucinations are images internally secreted by the brain. The conceptual avalanche caused by Penfield's glorified experiment on the cortex of conscious patients cannot be underestimated. Scholars and laypeople alike took his findings as the final proof that consciousness is generated inside the brain.[57] Countless movies and cartoons have been inspired by his experiments: *The Matrix, Inception, Inside Out*, and many others. Therefore, it might come as a surprise that the popular interpretation of his work is largely unsupported by empirical data. It is then worth going back to the actual empirical evidence to see where popular enthusiasm departed from facts.

From 1950 to 1958, Penfield collected a series of first-person reports about wakeful subjects' experiences during direct brain stimulation. He was neither first nor last in performing this kind of stimulation. Today, such invasive procedures, still applied in a restricted number of surgical procedures, have been superseded by harmless magnetic stimulation.[58] However, Penfield was the first to provide a detailed account of what subjects feel during direct brain stimulation. Briefly, things go as follows: one's brain is stimulated by an electrode and, as a result, one experiences something that is different from one's immediate surroundings. For instance, one sees flashes of light or one sees people who are not in the room. The popular interpretation is that one's experience is arbitrarily generated by direct stimulation and that *what one perceives is a movie generated and projected in the brain*. Surprisingly, though, such a conclusion is neither supported by Penfield's empirical evidence nor by subsequent cases of direct brain stimulation.

Contrary to popular wisdom, Penfield's direct brain stimulation never produced new experiences, but only reshuffled combinations of past events. Crucially, Penfield worked with adults whose brains were already entangled in a world of causal lines. Those brains were not isolated lumps of cells. They were parts of bodies that, in turn, were part of the world. With respect to the brain's power to create experience, the real experiment—which Penfield never performed—would have been the stimulation of a brain isolated since birth. Such a radical and never-performed experiment—akin to the scenario exploited by the sci-fi movie *The Matrix*—would be the only one that might really verify whether the brain can generate an inner mental world. Alternatively, as a second best scenario, Penfield might have looked for phenomenal experiences with little or no connection with one's actual life, something akin to extraordinary hallucinations. He did not. All patients reported only ordinary hallucinations whose content always consisted of reshuffled prior events, people, and objects. The outcome is that, contrary to a popular misinterpretation of Penfield's findings, the brain does not produce any pristine phenomenal content. The brain perceives one's past and, because of direct stimulation, reshuffles it. More precisely, its past produces effects in novel combinations.

Penfield was more like a plumber who, by tinkering with rusty pipes in an old building, succeeded in getting some water to spill out and wet the dusty floor. He did not create water by toying with pipes and valves. He did not create phenomenal experience by triggering neural firings. Electrical stimulation partakes of a larger historical causal network in which subjects' brains were entangled long before the moment they went under care. For instance, if a neural area has a certain structure because of an encounter with an enthralling partner, the encounter with the enthralling partner will be among the actual causes of any future activity taking place in that neural structure. Subsequent stimulations of one's brain partake of such a longer causal history. A certain activity in a certain portion of the brain is the causal outcome of past events as much as Penfield's electrode.

Surprisingly then, one of Penfield's most remarkable—and more consistent—findings is that hallucinatory experience is a reshuffling of past events. This finding has also been one of his most underestimated.

Empirical evidence shows that hallucinations are always and only "ordinary." English neurologist John Hughlings Jackson noted in his pioneering *Epilepsy* (1888) that, by and large, during hallucinations "old scenes revert." Penfield himself observed that hallucinations are always combinations of past episodes in one's life. It is informative to read a few excerpts from patients' hallucinatory reports as they were collected by Wilder Penfield and Phanor Perot in their monumental 1963 work on direct brain stimulation:

"Like company in the room. [...] it was like being in a dance hall, like standing in the doorway in a gymnasium like at the Kenwood High school. [...] People's voices." When asked, he said, "Relatives, my mother [...] It seemed as if my niece and nephew were visiting at my home. [...] They were getting ready to go home, putting their things on their coats and hats." When asked where, he said, "In the dining room – the front room-they were moving about. There were three of them and my mother was talking to them. She was rushed— in a hurry. I could not see them clearly or hear them clearly." (Case 2)

"After the visual sensation usually he would see a robber. Or a man with a gun, moving toward him. The man was someone he had seen in the movies or the comic strips. "Oh gosh! There they are, my brother is there. He is aiming an air rifle at me. [...] My mother was telling my aunt over the telephone to come up and visit us tonight. [...] My mother is telling my brother he has got his coat on backwards. I can just hear them." When asked if he remembered it, he said, "Oh yes, just before I came here." (Case 3)

"I hear music now—a funny little piece." Stimulation was continued. Patient became more talkative than usual, explaining that the music was something she had heard on the radio—that it was the theme song of a children's program. When asked, she said it was a record." (Case 6)

A voice saying, "Jimmy. Jimmy, Jimmy." When asked, she said it was her husband's name and what she calls him. (Case 8)

The surgeon said, "See if it comes again." Immediately upon application of the electrode, he said, "Yes, get out." It was a man's voice in a pool room, telling A. P. to get out. When asked, he said, "I have been in a pool room before; it might be that same one or it might not." Six weeks after the operation, in the clinic, A. P. was asked about this. He said that he remembered the incident clearly: it had happened about three years previously. He got angry after missing a shot at the pool table and broke his cue. The manager attempted to throw him out and a fight ensued. This incident made quite an impression on him. (Case 13)

I am sure there is no need to press the case any further. Penfield's findings never showed any evidence for anything but ordinary hallucinations—namely, that hallucination is a case of postponed and reshuffled perception. In all collected evidence, direct brain stimulation has never led to unexpected perceptual content. His experiments showed that direct brain stimulation allows patients to perceive the reshuffled past.

Hallucinations are a far cry from an arbitrary domain of unconstrained imagination as philosophers and laypeople usually think. What are dreams and hallucinations made of? *Like standard perception, hallucinations are made from the world, no matter whether the world is the one that just happened or the one that happened hours, days, months, or years ago.* Hallucinations are postponed perception of spatiotemporal composite objects.

• • •

A few more objections still stand. First and foremost, if we are identical with the perceived object when we perceive something and we are identical with the hallucinated object when we hallucinate, why do we not continuously perceive our past? In other words, why do we not hallucinate all the time? Conversely, what triggers a hallucination?

Hallucinations are taken to be mental phenomena because they are often triggered by events inside the head, such as when an electrode, a drug, and physiological processes seem to act inside the cortex. During hallucinations, the body often seems isolated from external causes. And the immediate causes of hallucinations, say, direct brain stimulation, do not seem to match one's experience. For instance, Penfield stimulates directly the subject's brain and, as a result, the subject hallucinates a piano concerto. Penfield's electric stimulus does not look like a piano concerto! Emily goes to sleep and, because of some internal process, she dreams she is walking on a solitary beach. The first reply to such objections is that one could apply the same rationale to standard perception as well. In fact, in everyday perception, too, the internal neural underpinnings do not match one's experience. Neural activity has both internal and external causes. When Emily listens to a piano concerto, her brain does not look like a piano concerto.

The solution is double preemption. A local cause—say, Penfield's electrode—preempts an already preempting cause that was blocking a remote cause—say, a piano concerto one attended to a few years earlier. The local cause switches off a blockade that prevents the perception of a past event. Empirical evidence supports this hypothesis. In fact, most alleged internal causes of hallucinations are disruptive processes rather than positive interventions. Many famous hallucinatory drugs—such as mescaline, ayahuasca, or LSD—decrease rather than increase brain activity.[59] Consistently, many drugs that increase brain activity—such as cocaine—are not particularly effective in producing hallucinations.[60] Occasionally, though, these drugs can nonetheless be disruptive and, thus, cause hallucinations. The inhibition of current perceptual stimuli is yet another case of double preemption. In fact, ongoing perception acts as an inhibiting factor of previous causal processes. Thus, depriving one of current stimuli allows more remote events to exert their influence.

The brain has a twofold function. On the one hand, it preserves the condition for future causal influx of events. On the other hand, it blocks such a causal influence by means of preempting causal structures. The preempting neural structure blocks other neural structures that would allow past events to produce effects now. Sometimes, an additional cause—an electrode, a drug, or a physiological process—interferes with this preempting neural structure, thereby blocking the blocker of the temporarily halted causal process. Imagine a water tap. Water would flow through the tap, but the tap has been turned off. The closed tap is a preempting cause of the water flow. Then you turn the tap on. The water flows. It is a case of double preemption. You prevent the water tap from preventing the water to flow. You do not pump water inside the pipes—you cancel a cause that prevents the water to flow. I consider here causal processes akin to the kind transference theorists have defended.[61]

In Figure 14, I sketch the preemptive model of hallucination. Normal arrowheads indicate causal relations, while round arrowheads indicate preemptive causal relations. An external cause—say, a red apple—causes a certain neural activity in one's brain. If there is no internal blockade, the red apple produces the neural activity. In standard perception, this is just what happens. It is worth stressing that the red apple is also re-sponsible for shaping the causal geometry of one's neural network in such a way that red apples trigger that particular neural activity. As a result, when there are no apples nearby, the neural activity takes place because of the original cause, namely the first red apple and then the following ones. Yet, in normal circumstances, the causal flow is pre-empted by some neural blockade, which is internal to the brain. The neural blockade is like a water tap: it stops causal processes from flowing and thus producing causal effects. However, when another event—whether internal like sleep or external like Penfield's electrode—pre-vents the neural blockade from preventing the remote cause, the first red apple will exert its causal influence on the brain. Such an event is a double preempting cause: it is like someone turning off the water tap.

In fact, hallucinations often arise whenever an inhibiting factor is switched off. This is consistent with the fact that hallucinations arise either by depressing current perceptual stimulations or by disrupting inhibiting factors.[62]

The preemptive model of hallucination helps to explain cases such as Penfield's electric stimulation in which a disruptive event, such as an electric discharge, disrupts normal neural activity and causes a hal-lucination. Likewise, isolation, sensory deprivation, and sleep induce hallucinations by diminishing the strength of neural inhibiting mech-anisms. The preemptive model does not need any cumbersome inter-nal representation.

Hallucinations are not the reactivation of stored content. They are not a movie in the head. They are tantamount to the opening of a causal water tap. Penfield's electrode is not the carrier of what his pa-tients experience. Neither does it revive the mythical "engram" that Penfield himself adopted as a possible carrier of content, "there is more

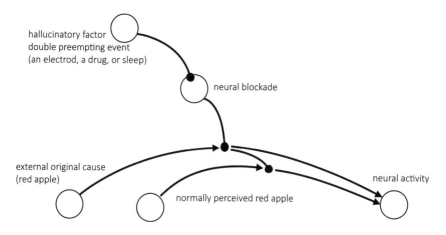

Figure 14. Hallucinations as double preemption.

than that to the special mechanism of the mind. Its action accompanies consciousness and, in it, a complete record or engram is preserved." No engram lurks amidst neural connections like an updated—and scientifically disguised—version of Descartes's impression and ideas.

Modelling the hallucinatory factor as a causal enabler—as a causal factor inhibiting another causal inhibitor from preempting a causal chain from occurring—has two main advantages. Here, I borrow Jonathan Shaffer's terminology. First, it explains the lack of specificity between the hallucinatory factor and the resulting hallucination. Second, it explains hallucinations in terms of standard perception. The theory of spread mind suggests the existence of a physical object for every hallucination. If I hallucinate a red apple, the neural activity in my brain is due to the red apple that I saw and I still perceive, albeit in a delay. The non-specific hallucinatory factor removes a causal blockade that would otherwise prevent the causal process to reach completion. The preemptive model of hallucination sets aside the traditional notion that the causes of one's hallucinations are solely inside one's head. Hallucinations and dreams are explained as cases of reshuffled delayed perception. Neither do they access an inner private mental world nor do they rehearse internal representations.

A hallucination is a bit like a flow of vehicles trying to get past a busy traffic circle or roundabout. Normally, the incoming flow of causal vehicles from the temporal and spatial surroundings is so intense that it prevents any other vehicle from getting into the roundabout. However, if such a flow is either interrupted or severely disrupted, other cars—akin to more remote causes—have a chance to get through the roundabout. It is not by chance that most forms of experience that are not caused by a proximal object are the side effect of the decrease of other sensory modalities. Dreams require cognitive, psychological, physiological, and environmental isolation. Similarly, Charles Bonnet syndrome, which is a hallucinatory condition, is caused by the disruption of a whole cortical input area.[63]

Sensory deprivation often determines hallucinations by diminishing the influence of closer events.[64] Many forms of drug-induced hallucinations are the consequence of inhibiting rather than excitatory factors.[65] Dreams too occur thanks to mechanisms that shut down or considerably reduce the influence of immediate stimulation. Finally, the mechanism leading to various form of hallucinations, dreams, hypnopompic, and hypnagogic images is the imbalance between the influences of events that took place at different times.

The presented model of hallucination dwells crucially on inhibiting factors. The activity going on in the brain is always the result of the past. However, for many practical reasons, while the brain is the result of all its causal history, at any time only a few of these causes are allowed to produce effects. For instance, you dodge because you see a menacing hornet flying toward you. The dodging is the effect of the hornet that, at that instant, is the main cause that propagates its influence through your body. Eventually, you daydream of your past holidays in Tuscany and, in that moment, the causal structure of your brain is taken over by those serene days. During the ensuing night and sleep, the presynaptic inhibition of the afferent endings of sensory nerves causes a relative sensory isolation of the brain. In this way, the influence of the proximal external world is blocked.

To sum up, hallucination-inducing conditions—such as daydreaming, dreams, phantom limbs, Charles Bonnet syndrome, and many others—have three factors in common. First, one hallucinates objects or properties of objects that one has perceived before. Second, hallucinations are more likely when the causal influx from one's surroundings is somewhat reduced. Finally, hallucinations are the offshoot of neural disruptive factors. To recap, all these conditions are cases of ordinary hallucinations.

The first factor has been addressed. It is key since the hallucinatory content is not an arbitrary mental experience concocted inside the head, but can always be traced back to the external world. Hallucinations are a form of perception. The last two factors are linked together. They hint to the cognitive and neural underpinnings of hallucinations. Whenever normal causality is disrupted, the influence of remote events is greater.

This account does not address the selection mechanism of the events one hallucinates or dreams. Neither does it address the apparently significant organization that dreams often show. This account addresses a preliminary but fundamental issue about dreams and hallucinations, namely why and what one experiences. Both questions are not addressed in neuroscience and in psychology. Once these questions are addressed, various mechanisms can be taken into account to explain why a particular selection of past events is active. The theory of spread mind does not address why one dreams a particular combination of objects and events, but rather why one experiences something at all.

.   .   .

Suppose that, as in Penfield's findings, all cases of hallucinations are rooted in causal commerce with a physical object and thus, more bravely, that each experience is identical with an object. When we hallucinate a previous object, we have a postponed perception. But how can we perceive new combinations of objects and properties that we have never perceived? How can we dream of a flying pink elephant?

The answer—already partially outlined—is that sometimes perception proceeds through multiple causal pathways of different lengths, picking up and singling out spatiotemporally composite objects. The flying pink elephant is akin to looking at a set of scattered objects reflected by a shattered mirror. The elephant is not different from the beloved red apple. They are both bundles of various physical parts and

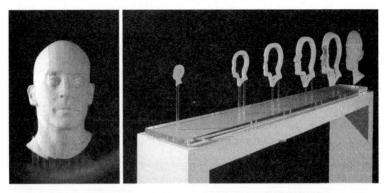

Figure 15. Jonty Hurvitz's *Hurvitz Singularity* is a spatially composite object.

Figure 16. Bernard Pras's *Dalí.* Anamorphic sculpture as composite objects.

properties. I am purposefully vague on the ontology of the ultimates that partake of an object occurrence—whether they are properties, tropes, parts, or whatever. As far as I am concerned, the only requirement is that such constituents play a causal role—i.e., they are physical. The difference between the red apple and the flying pink elephant is a matter of degree, not an ontological gap. The red apple is more easily perceivable and controllable. The flying pink elephant requires a more articulate spatiotemporal carving. Besides, it is easier to envisage the causal structure involved in the perception of the apple than it is in the case of the flying pink elephant.

The case of the red apple is simpler since its constituents—the red peel, the spherical mass of pulp—are in the same location at the same time. They produce a joint causal effect much more easily than the spatiotemporally scattered flying pink elephant. In practice, but only contingently, most visual systems allow the red apple to exist. Assembling the flying pink elephant is more difficult since it requires, at least, three distinct causal pathways picking up three different constituents from one's lifetime: something pink, something flying, and an elephant. Yet, as the example of false mirrors shows, sometimes a perception takes place in unusual ways. Multiple causal pathways allow spatiotemporally scattered elements to act together.

Everyday perception is constrained by the spatial and temporal order of objects and their properties.[66] Only red apples and the like are perceived. Yet, a partially transparent mirror or a kaleidoscope can make a difference in the causal geometry of the world. A neural system causally coupled with gazillions of events can make a bigger difference.

An object is an actual cause thanks to two factors: on the side of the constituents and on the side of the perceiver's body. A few examples will help. A brass sphere acts as a sphere with little help from the surroundings circumstances. A set of air pressure waves composing Leporello's aria requires a complex auditory system connected with a sophisticated cognitive system. Contrast a painting on a wall with a heavily encrypted and compressed image of the same painting scattered across multiple files over a distributed network. The former is easily assembled by any visual system capable of singling out features grouped on a canvas. The latter requires a complex set of algorithms singling out physically wildly scattered records. Yet, once they interact with the proper combinations of physical structures, they both take place as an actual cause—no matter how complex is the necessary physical structure with which they interact.

In short, there is no ontological difference between the flying pink elephant and the red apple on my table. The difference is practical. It does not dwell on the boundary between mental and physical. Rather it is based on the extent of the causal work that a body has to do in order to interact with a given object. The three crosses seen previously exemplify the same point (Figure 4). A continuum of cases can be envisaged.

At the lower level, consider a static object like the red apple on the table. No matter how I look at it, the object is sturdy, still, stable and, to a great extent, invariant to any change of perspective. It is approximately a red sphere. Most familiar objects have the same attitude. Furthermore, such objects react consistently to movement. In fact, if I move a part of them, all their parts will move together as a whole—e.g., if I push a leg of a table, the table will move as a whole. Occasionally, this is not true. When cohesion does not occur, we are uncertain about the unity of the object. Nevertheless, several conditions are a sort of rule of thumb for singling out objects, such as pushing and moving together, hanging together, sameness of texture, color, and the like. A temporally distributed object like a tune played by an orchestra has a more elusive causal existence but it is nonetheless treated like a whole by most people. A tune can be listened to, sold, and recorded. Causal processes carve objects out of the physical continuum.

All objects are combinations of spatiotemporal parts, be they flying pink elephants or red apples. We refer to some of them as objects and to others as hallucinations. Yet, they do not differ in kind.

The spatiotemporally composite nature of reality has often been exploited in visual arts. Artists have been fascinated by the possibility of visualizing the spatiotemporal scattered structure of objects. Historically, classic examples range from analytic cubism to the omnipresent Dali portrait of Voltaire. The baroque painter Arcimboldo is another excellent example. To a large extent, pointillist painters such as Signac or Seurat addressed the same issue from a sensorial perspective. Consider Jonty Hurvitz's *Hurvitz Singularity* and Bernard Pras's anamorphic sculptures. These artworks exemplify the composite nature of objects by showing in an explicit way the many different objects placed in a spatiotemporally spread region. Objects do not exist by themselves. Taking place as a whole requires coupling between certain stuff in the environment and other physical systems, in this case a human body. Such artworks embed the constituting causal process.

For instance, consider the *Hurvitz Singularity* (Figure 15). Such an object is a head only when it is seen from a specific standpoint. However, it is not a "mental head." It takes place as a head only from a specific angle. With respect to a normal head, it requires a greater commitment from the side of a perceiver who has to be in a unique position in space. While the red apple exists thanks to processes that might happen from every angle, Hurvitz's singularity exists only from one specific viewpoint.

A temporal refinement can be envisaged. The slices that *Hurvitz's Singularity* is made of might be placed at a greater distance. A mechanism that makes them appear and disappear at fixed intervals might be introduced. In this way, the whole head will exist only if a human body is at the right place and at the right time. Nevertheless, the head is neither a hallucination nor a mental ghost. The head is an object whose existence is constrained by causal, temporal, and spatial factors. It is a spatiotemporally composite object. Traditional heads are objects whose existence is similarly constrained. To some extent, *Hurvitz's Singularity* is halfway between a standard head and the hallucination of a head.

Even more striking is Bernard Pras's sculpture *Dalí* (Figure 16). If looked at from a proper viewpoint, the scattered group of objects will become Salvador Dalí's head. However, if we look at the group of objects from another perspective, they are only a bundle of scattered objects spread in space and time. Yet, Dalí's head is not a mental concoction. It can be shot by a digital camera. When we look at the sculpture from the right perspective, we see a real head. The interplay between objects and causal processes is brought to the fore by the ingenuity of the sculpture. Pras's sculpture, too, is a nice model of both perception and hallucination.

Consider once more one of my favorite examples, namely, a constellation. The constellation is akin to both Hurvitz's and Pras's sculptures. It is an astronomical object and it is indeed a case of perception. The stars that compose a constellation are akin to Hurvitz's slices or to Pras's scattered objects. Such stars are placed at different locations and, more astonishingly, at different times. The constellation as a whole is spread both in space and time. Still, it is not a mental object. It is a physical object that we perceive, the same way that we perceive everything else. It is a physical object located where and when its elements are, but it exists only relative to a certain space time location—i.e., where the earth is.

Now, consider the amount of causal coupling that neural and sensory systems are responsible for. When I use the words "neural" or "sensory" or "perceiver," I do not attribute a special status to these physical structures. Simply, I refer anatomically to a part of the physical world that happens to have a causal role in allowing certain causes to take place. Events taking place in different places and times in one's life produce joint effects in any order they like as long as they interact with the proper physical structures.

. . .

In the philosophy of perception, the argument from hallucination has endorsed both the downgrading of perception to a reliable hallucinatory state and the separation between appearance and reality. Yet, such an argument is of an empirical nature: either empirical evidence supports it or its strength wanes. So far, this chapter has strived to show that such supposed empirical foundations are indeed questionable. My claim is that no empirical evidence backs up the separation between experience and the world once the evidence is not interpreted based on the very conclusion under scrutiny. In this regard, a traditional issue that has held center stage in the current philosophical debate about perception and hallucinations—the common kind assumption—has been used to argue that perception is a kind of hallucination. In contrast, I will argue that the common kind assumption entails that hallucination is a kind of perception.

In brief, the common kind assumption is conveniently split in two as the Scottish philosopher Fiona Macpherson clearly states in her 2013 book on hallucinations:

- In hallucination no real object is there
- Introspectively, hallucination and perception are the same

Both points are of an empirical nature. They are not conceptual truths. They are empirical facts that have to be ascertained in the world. Anyway, based on such premises, a spreading strategy is developed routinely. The standard spreading strategy argues that, since no difference distinguishes perception from hallucination and since during hallucination there is no external object, in perception, one experiences something that is not an external object. Apparently, the common kind assumption—i.e., the two premises together—delivers a mortal blow to any hope for realism. Realists, though, counterattack by rejecting the second premise. They deny that perceiving a dagger and hallucinating a dagger are the same. The two cases appear to be the same, but they are not. Hallucination and perception are different but, due to cognitive shortcomings, one cannot report the difference. It is the disjunctivist move—i.e., the difference exists but we are not aware of it. The theory of spread mind offers a solution that saves both realism and introspection.

Rather than denying the second premise, I reject the first one: namely that when I hallucinate, no physical object exists. Contrary to received wisdom but coherently with empirical evidence, when I hallucinate, a physical object is always available. Thus, it is not true that when one hallucinates no external object is there. A spatiotemporally composite object is always available. The notion of *there*, its scope and extension, is stretched to encompass all one's life.

The first premise of the common kind assumption is false—*empirically* false. Thus, the second premise is true but it means the opposite of what dualists and sensationalists think. Consequently, the theory of spread mind endorses realism and physicalism (in their strongest versions) but it sets aside disjunctivism. The model presented in this chapter is compatible with other views, most notably with various kinds of direct realism or relationalism. The theory shows how to locate a real physical object for all cases of experience—for hallucinations, dreams, and so forth. Thus, realism is saved without having to discard introspection—*hallucination and perception look the same because they are the same.*

Indeed, in the presented account, no ontological difference distinguishes hallucinating a dagger from seeing a dagger—in both cases there is a dagger. In everyday perception, the dagger is a few meters and a few hundreds of milliseconds away. In a hallucination, the dagger is many hours, days, or years away. However, in both cases a real dagger is the cause of one's neural activity. One might rebuke the idea that—in the case of hallucinations—there *was* a dagger. Yet, in everyday perception, too, there *was* a dagger. In everyday perception, the dagger one sees is separated only by an imperceptible fraction of an instant. Yet, the dagger *is* in the past, too. The difference between hallucination and perception is only quantitative.

The direction of the spreading strategy, normally moving from hallucination to perception, is reversed. Now, the explanatory strategy goes from perception to hallucination. Hallucinations and the like are forms of extended perception. Realism is safe!

The conceptual effect of such an inversion—explaining hallucination in terms of perception rather than perception in terms of reliable hallucination—cannot be underestimated. Since Descartes, the ancillary role of perception has held hostage most scientific and philosophical reflections. Scientists, philosophers, and laypeople alike have believed in the difference between what the world is and how the world appears to be. As a result, the notion that we live in a kind of perennial dream or hallucination, which our senses and brain painstakingly keep in synch with the external world, has dominated the western Zeitgeist for centuries. Setting aside the argument from hallucination paves the way to eradicating any distinction between experience and things, between appearance and reality, between mind and world. The rejection of the first premise of the common kind assumption (the lack of an object) allows us to set aside a whole cohort of ancillary notions such as semantics, supervenience, and ontological levels.

It is well known that the argument from hallucination gives support to the notion of local supervenience that, in turn, makes plausible the argument from hallucination. Once the argument of hallucination is set aside, though, what proof do we have that our experience is locally supervenient on brain activity? None, because, at some point in the spatiotemporal manifold, the external object occurs. Of course, if one assumes something akin to strong local supervenience, hallucination will supervene on the state of the subject's brain. Yet, assuming local supervenience begs the question because one assumes what one should prove, namely that hallucinations require no more than certain states of the brain. Remarkably, the empirical evidence, contrary to entrenched views, points in the opposite direction—no cases of world-unrelated hallucinations have ever occurred.

6.

A Zoo of Objects and Experiences

> The title [...] would justify the inclusion of
> Prince Hamlet, the point, the line, the place, the
> hypercube, all generic nouns, and perhaps,
> each one of us and the divinity as well.
>
> —JORGE LUIS BORGES, 1969

THE KEY CLAIM is that hallucinations are *always* postponed reshuffled perception—i.e., ordinary hallucinations. Yet, some readers might disagree, pointing to alleged cases of hallucinations with no contact with one's actual world. They might appeal to the existence of these extraordinary hallucinations—hallucinations of objects one has never perceived—as deadly counterexamples to the theory of spread mind. My reply to such an objection is that such cases are nothing but a scientific myth.

It is not rare to come across claims about extraordinary hallucinations such as colored dreams in congenitally blind subjects, innate corporeal feelings, phantasmagoric hallucinations, and so forth. The significance of such cases depends on their empirical soundness. In truth, has anyone ever hallucinated anything that was not part of one's world? My contention is that no human being has ever experienced something that has not been part of their actual physical world.

On empirical grounds, in this chapter I survey a relatively long list of exotic cases that have backed up the claim that experience is cooked up inside the brain. The picture that will emerge is radically different from the received view. Evidence does not support the autonomy of experience. On the contrary, evidence supports a complete match between appearance and reality, between mind and nature, between experience and world. Thought-experiments—such as zombies and brains in a vat—are here ignored since they restate prejudices derived from supposed yet non-existent empirical evidence. Thought experiments are useful to measure the extent of our ignorance rather than to state something about the world. They reveal conceptual loopholes, at their best. First and foremost, I want to revise empirical evidence. If such evidence is missing, many traditionally impressive and respected thought experiments, arguments, beliefs, and soothing allegories will no longer be convincing.

We will wander through a zoo of cases in which one's experience does not seem to be strictly related to one's external world, and we will see that whenever one experiences something, a corresponding external physical object exists.

. . .

The first notion I would like to address and debunk is the belief in the existence of purely endogenous mental experience—namely, something our brain should create out of thin air. It is surprising that such a conception is so popular when no evidence backs it up. If my gaze wanders across my everyday life, all my experience is made of objects, people, and their properties. When I dream, too, my dreams are made of objects, people, and their properties. Yet, the scientific literature has often implied that one's experience can be autonomous from the physical reality, a claim often based on a limited number of exotic and problematic reports. Perhaps inadvertently, science has fed the myth of endogenous mental experience—namely, that experience might not match the physical world.

Occasionally, a group of scientific and philosophical myths easily gains common acceptance. In our case, most people think the brain "cooks up" colors, sounds, faces, people, phantasmagoric images, and so forth. Such myths are taken to be true, because they confirm the very theoretical expectations from which they spawned. I here think of what a great 19th century Italian writer, Alessandro Manzoni, wrote regarding the relation between scholars and the expectations they rise: "It cannot be expressed how much authority the opinion of a learned man by profession carries with it, while he is attempting to prove to others things of which they are already convinced."

Yet has any subject ever perceived or experienced a pure mental concoction? If we examine all reports and all cases punctually, we will discover that whenever one experiences something, it is always possible to trace back that particular experience to a physical entity. In other words, extraordinary hallucinations are an empty class. All known hallucinations boil down to cases of ordinary hallucinations. There are good reasons to suspect that most evidence has been misinterpreted and that the existing data has been overestimated, all in the service of supporting the myth of endogenous pristine mental experience. In some cases, the faith in such a myth is so strong that, in the absence of actual data, scholars often assume that if evidence were available, it would support the existence of extraordinary hallucinations. The only problem is that such evidence is missing. In contrast with such an attitude, the empirical evidence shows that all human experience matches the physical world.

The myth of endogenous mental experience reminds me of the myth of spontaneous generation—namely the idea that life forms should be able to autonomously create different life forms like maggots, fleas, and worms. In fact, it has been observed that if a piece of meat is left unattended, other life forms are soon thriving inside it. Louis Pasteur and, much earlier, Francesco Redi showed convincingly that a completely isolated piece of flesh is unable to produce other life forms. Likewise, people believe commonly that a brain in a vat can autonomously generate mental experiences just like a piece of flesh under a glass bell was believed to autonomously generate life forms. In both cases, the lack of understanding of the nature of the phenomenon—life or consciousness—has led thinkers to postulate an unknown generative principle inside a biological structure.

As mentioned, if one thoroughly peruses reports from dreams, one will only find the actual world, variously reshuffled and reorganized.

No purely mental entity will show up. One will find only a kaleido-scopic recombination of actual objects, individuals, and their proper-ties. No unexpected colors, mysterious tastes, extrasensorial experiences, or Lovecraftian geometries populate people's dreams. Real cases and real reports are much more mundane. We saw the case of Penfield's patients. In the following, I extend the survey. My main line of attack will always be the same: for every alleged case of endog-enous mental experience, I point to a physical object in one's life. Poi-gnantly but coherently with the theory of spread mind, the alleged cases of pure mental experience are few, often ill-documented, and likely to be the result of a biased interpretation. I would like to list the most well-known cases with their empirical shortcomings:[67]

- congenitally total blind subjects allegedly capable of dreaming or imagining colors have been based on a handful of cases for which there are conceivable alternative explanations;
- phantom limb syndrome in congenitally limbless patients is based on an extremely limited number of cases and there is latitude for available alternative interpretation;
- phantom penises in pre-operated female-to-male transsexuals are based on a very limited number of cases with possible al-ternative psychological causes;
- phosphenes are likely delayed perception, with available alter-native interpretation;
- impossible hues such as reddish-green or yellow-bluish are explainable as extended perception of more physical colors;
- super-saturated colors are likely a wrong interpretation of data;
- afterimages and aftereffects can be explained in simpler terms;
- Hume's missing shade of blue is an empirically unverifiable claim;
- color-synesthesia in congenitally either color-blind or blind subjects has never been documented;

▪ visual aura in congenitally blind subjects has been occasionally mentioned but never reported;

▪ having totally new phenomenal experiences either in dreams or in other circumstances is beyond experimental validation.

Before examining such a repertoire of bizarre cases in detail, it is worth stressing that the overwhelming majority of cases of perception, dreams, hallucinations, illusions, and pathological syndromes stand in stark contrast to these outliers. These cases are extremely rare situations that have been painstakingly chased by philosophers and neuroscientists alike in order to find confirmation for the capability of the brain to create a mental world. Yet, they are so rare and light on details that, given the weight of proof that they should bear, one should be quite skeptical about them. An extraordinary claim—such as the existence of phenomenal properties of experience—requires an extraordinary proof. These cases are not even close. Yet, they have often been accepted unquestioningly because they confirm widespread and popular prejudices. If experiential content were really concocted by brain activity, one ought to expect that, in cases of disease or dysfunction, deviation from the actual world should be the norm rather than a rare exception. This is not the case.

· · ·

A familiar case of alleged purely mental or neural experience is offered by phosphenes. I start with them both because we have all experienced them and because they are not obviously related to physical phenomena. If one gently presses one's eyeballs, after a few seconds, geometric patterns will be seen. The experiment can be repeated in a completely dark room. Thus, one can wonder whether such an experience corresponds to a physical phenomenon or if it is a purely mental event.

Phenomenologically, phosphenes are often described as elementary visual experiences produced by nonvisual local stimulation of the visual system. On the one hand, we can categorize phosphenes based on the kind of physical phenomenon that is used to elicit them—mechanical, chemical, magnetic, or electrical. More rarely phosphenes can be elicited "by the energetic radiation from lightning flashes and thunderstorms."[68] On the other hand, phosphenes are classified either as cortical or retinal based on the location where the stimulation is applied. The vexing question as to whether cortical stimulation activates the retina by means of a possible conduction is of no interest here.[69] In general, philosophers like phosphenes because they are taken to be solid evidence that visual experience can be "internally generated." In this regard, Ned Block is confident that, "We are now able to produce very simple visual sensations such as the illusion of the presence of flashes of light (phosphenes) by means of direct neural stimulation." Yet, as we will see, direct neural stimulation is not sufficient—some previous causal entanglement with light is necessary. In fact, the theory of spread mind provides an alternative account of phosphenes. Since light is the most common cause of the activity in our retina and thus in our cortical areas, a sighted subject has very easily postponed perception of elementary light objects. Thus, phosphenes are hallucinations of elementary light objects—i.e., postponed perception of light.

In normal conditions, light hits the photoreceptors and elicits effects in the cortex. Such effects single out light-related events in the world. Eventually, another physical cause (for instance, applying pressure on one's eyeball) creates the same activity both in the retina and in the corresponding neural stages. Such an activity takes place only because, in the past, light shaped one's neural pathways. Light was among the critical causes responsible for the development of one's neural pathways, from the retina up to the cortex. In fact, light is the original cause and it is still part of the causal history.

What do I see when I see a visual phosphene because of a pressure on my eyeballs? I see light. Light is the original cause of that retinal activity. A local pressure enables past lights to be, once again, causes of occurring events. The hallucination I see is akin to those produced by Penfield's electrodes in the brain. The difference is that phosphenes are much simpler occurrences and can be triggered by coarser and simpler causes either in the periphery of one's body or in the cortex. According to this account, if one had never had any contact with light, one could not see any light by means of pressure on the eyeballs. On the other hand, if neuroscientists succeed in connecting a congenitally blind subject's optical nerve (or any afferent nerve) to an eye (either biologically or artificially), the resulting experience would be of visual character. This is indeed what happens with modern visual prostheses.[70] In fact, such prostheses establish a causal link with light.

If the above conclusion is correct, why do we see visual phosphenes when no visual phenomena are around? Why do we see anything at all when non-visual causes trigger our peripheral nerves or we bump our heads? We see light because these additional causes—pressure, strokes, electric and magnetic pulses—preempt neural blockades between pre-existing causes and the neural activity. In normal sighted subjects, both the optical nerve and the striate cortex have developed because of the causal influx of optical phenomena. As a result, the causal outcome of optical phenomena takes place. *A phosphene—either cortical or retinal—is a little hallucination triggered by an occasional cause.* The occasional cause, as in the more complex episodic hallucinations addressed before, does not trigger the activation of internal representations. The occasional cause temporarily inhibits the mechanism that prevents one from being overwhelmed by one's past.

Likewise, the diffused red glow that one sees when closing one's eyelids before going to sleep in a dark room can be explained. Such a glow is the persisting causal coupling with various light waves that one has been in contact with for the best part of the day. As long as the eyelids are closed, no stimulus is exerting any causal influence. A minimal sensory deprivation is achieved, a sort of toy version of Charles Bonnet syndrome. Thus, one keeps seeing the last causes of neural activity. The causal influence of the last cause persists in lack of more pressing ones.

The theory of spread mind puts forward two predictions as regards phosphenes obtained by means of physical pressure. The first prediction is that, if such pressure were maintained for a long enough period and if one's neural structure were adaptable enough, one should expect a progressive reduction in the visual nature of experience and an increase in the tactile nature. The second prediction is that, if a subject had had no contact whatsoever with light since birth, the subject would not have any visual phosphenes. The subject, though, would use the pressure-induced release of rhodopsin by cones and rods to single out pressure-related phenomena. In other words, the subject would use the eyes as glorified pressure sensors. The subject would have an experience of tactile events. Curiously enough, there is an intriguing report of an early case of eye surgery in which, after several years of relatively complete blindness, a subject reported, following sight restoration, tactile sensations induced by light. The case was reported in 1728 by the surgeon William Cheselden: "When he first saw, he thought all objects whatever touch'd his eyes, (as he express'd it) as what he felt, did his skin."

Thus, phosphenes can be accounted for in terms of hallucinations and the memory of elementary light objects—namely, the postponed perception of light. In this respect, they are not evidence of an inner mental world. They can be traced back to light the same way as everyday perception.

However, a different matter would be the case if a congenitally totally blind subject reported visual phosphenes. Such a case would be key because it would be—in regard to visual experience—a reasonable approximation of a brain in a vat since birth. Such a case, though, has never been reported.

Due to various misunderstandings, many scholars and most laypeople believe that congenitally blind subjects report visual phosphenes when their occipital cortex is triggered.[71] This is not the case. There is not a single report of congenitally blind subjects experiencing phosphenes. Such a misconception stems from two common sources of confusion in the scientific literature. First, totally blind subjects are not necessarily congenitally blind and congenitally blind subjects are not totally blind. Second, subjects are often classified as congenitally blind because of the early onset of their condition. Because of the time course of their condition, such subjects have the opportunity to experience light-related phenomena for a relatively long span of time, often several years.

Many patients qualify as blind subjects only because they are not able to use their visual system—for instance, they can neither recognize objects nor single out shapes. Yet, phenomenologically, they are not blind. Even more questionably, since blind subjects are relatively rare, they are often mixed with normally sighted patients that do, of course, report phosphenes. Unfortunately, most of the literature is oblivious to whether the patients are congenitally blind or not, let alone whether they are completely blind.[72]1

In this regard, in 1968 the neurosurgeons Giles Brindley and William Lewin wrote a seminal paper, *The sensations produced by electrical stimulation of the visual cortex*, that spread the notion that blind subjects could see light by brain stimulation. Surprisingly, their patient was not congenitally blind. Actually,

> [the] patient, aged 52 years, who had been myopic from childhood, developed bilateral glaucoma in 1962. Vision failed progressively and then in 1967 after a right retinal detachment she was left blind, despite several corrective operations. The patient could only recognize a flash of light in a narrow strip of the temporal field of the right eye, and hand movements in a small part of the peripheral lower temporal field of the left eye.

So the patient was not blind until she was in her forties and even afterwards she was still able to recognize a flash of light and movements of the hands. Such skills are a far cry from being phenomenologically totally blind. Likewise, none of Penfield's patients was blind.[73]

In 2000, in another often-quoted paper, *Magnetically induced phosphenes in sighted, blind and blindsighted observers*, the neuroscientist Alan Cowey and his group considered eight patients, six sighted and two blind subjects (PS, age 61, blind when 53 years old; GY, partially blind due to "almost complete destruction of left striate cortex V1, with some additional damage to extrastriate areas V2 and V3" when 8 years old). Meaning that none were completely congenitally blind.

Finally, another classic paper is *Effects of a Hallucinogenic Agent in Totally Blind Subjects*, by Alex Krill and his team in 1963, about the capability of congenitally blind subjects having visual hallucinations by means of hallucinogenic drugs. At the very onset, in the method section, the authors state that (italics mine):

> Twenty-four totally blind subjects were studied. Four subjects were totally blind by the age of 2 and were designated as "congenitally blind." *All of these subjects probably had some vision at birth.*

Thus, none of them were congenitally blind. Note also that the alleged totally blind subjects were a small percentage of the total involved in the study (four out of twenty-four).

These examples should be enough to induce us to be cautious about the alleged evidence of phosphenes in patients lacking contact with light. In fact, in this regard, the neuroscientist Janna Gothe wrote that "The ability to elicit phosphenes is reduced in subjects with a high degree of visual deafferentation, especially in those *without previous visual experience.*" Indeed, the more one checks the data thoroughly, the more it seems that the ability to elicit phosphenes is absent in the congenitally blind.[74]

A conclusion that can be drawn from available empirical evidence is that no completely congenitally blind patient has ever reported visual phosphenes. No evidence supports the claim that the stimulation of the central nervous system is sufficient to elicit phosphenes. It is, of course, a different matter if a brain has previously had contact with light.

•   •   •

The saga of the congenitally blind does not end here, though. Regardless of phosphenes, the capability of congenitally blind subjects to concoct visual experiences is a never-ending myth. The myth is so strong because it supports the autonomy of our alleged mental world. Unfortunately, it is not the case. In this section we address the alleged evidence of the capacity to concoct visual images and colors by sheer mental power. I will focus on whether a congenitally blind person can concoct a visual experience.

As seen, in the literature about blind people there is some ambiguity between sighted subjects, the partially blind, the blind, the congenitally blind, and the congenitally totally blind. Only the last group is relevant as to the existence of inner phenomenal experience and it is always, to the best of my knowledge, empty. The bottom line is, to the best of my knowledge, that the patients who are *congenitally totally blind* report *no visual experience of any kind*. Let us consider three kinds of visual experience—shape, color, and depth (perspective).

Consider shape first. Since blind subjects deal with shapes and thus experience them, a widespread belief regards their alleged capability to have mental visual images of forms. Such a belief is misplaced. Congenitally blind subjects are capable of direct physical contact with the world. They touch objects and move in space. Consequently, they are capable of experiencing shape, size, movements, speed, and locations. They do not do this visually, but grasp physical properties such as shape and size by means of a collection of other causal pathways, as reported by Lochlan E. Magee and John M. Kennedy in 1980. Sighted subjects do the same, but are overwhelmed by the richness of vision. Thus, congenitally blind subjects who have phenomenal experiences made of shapes do not provide evidence that such subjects have any visual experience. They have, say, shape-experience, which is not visual experience. For instance, they experience a circle through touch. Although many believe that a circle is a visual notion, it is not. We can see a circle, but the circle is not necessarily a visual form. It is a shape. We can touch a circle. In many respects, the circle is more easily perceived through touch than through vision.

Colors are peculiar to vision since a person cannot be causally coupled to colors without sight. Colors cannot be perceived through touch. In fact, while shape is a common constituent of congenitally blind subjects' experiential world, color is absent. Such an asymmetry should raise serious doubts to fervent supporters of the capability of blind subjects to create mental experience with a visual character. The difference between the two cases is easily explained by appealing to the fact that shape can be grasped through touch while colors are only physically accessible by means of sight.

Regarding color mental imagery and blindness, it is worth addressing the unique case of a subject reporting to be both congenitally blind and yet capable of mental awareness of color. This is the Turkish painter Esref Armagan, who allegedly reports color mental imagery.[75]

Here, we should distinguish the painter's artistic skills from the controversial claims about his color visual imagery. While he paints complex scenes using various perspective-related techniques, how can we validate his claims about color mental imagery? It is well known that blind subjects are aware of color associations to physical phenomena—such as references to the blue of the sky or to the red of blood—however, no evidence shows that congenitally blind subjects experience colors. It is unclear to what extent a congenitally blind subject's references to colors are the result of learning or the result of some innate or *a priori* skill. In the case of Armagan, he depends on sighted subjects to associate colors with objects. Reasonably, he must depend on sighted subjects for painting using colors. At no time can he see the colors he is handling. Consider the following cruel trick. If one replaced the color tubes with otherwise identical black ones, Armagan would have no way to know which tube he is using. His artwork would then result in an indistinct black surface. Nothing short of a miracle would allow him to behave otherwise. His skill, therefore, seems more of a cognitive ability than the outcome of an inner mental world.

What about perspective? Once again, blind subjects can learn to master vision-inspired graphic techniques. Perspective is one such example—the ability to grasp the structure of perspective is not necessarily a visual skill although it is commonly exploited by our visual system and by visual techniques of representation. Although perspective is normally absent from blind subject's drawings, blind subjects can nonetheless understand the principles of perspective and learn how to draw according to such conventions.[76] Yet this fact is no proof that they see perspectival images with an inner mental eye. It only shows that perspective, like any conceptual set of conventions, can be learned and mastered by subjects regardless of one's sensory capability. After all, it is well known that in the course of history, artists achieved perspective only very late and after a lot of theoretical experimentation. In the 1970s, James J. Gibson observed that the late appearance of perspective in artworks is the proof that human subjects do not see an internal pictorial image. If we had access to an internal perspective representation of the world, it is conceivable that we might draw perspective images more easily. This does not seem to be the case. A congenitally blind subject preserves the absolute shape of an object like ancient artists did, rather than bending angles and shrinking shapes based on their relative distance and position with the observer. Of course, if congenitally blind subjects are trained they can learn perspective-related rules and apply them. Similarly, too, we can learn all kinds of geometric transformations without having any phenomenal epiphany.

Blind subjects obtain perspective skills because of the intervention and guidance of sighted peers, notwithstanding frequent claims to the contrary. The psychologist Fernando Lopes da Silva at the University of Amsterdam is working with congenitally blind subjects. This is a description of how a blind subject appreciates a perspective landscape rendering he outlines in the article *Visual Dreams in the Congenitally Blind?*, published in 2003.

> Were a blind man standing at the Place de la Concorde asked to trace with his hands each side of the Champs Elysées to the Arc de Triomphe, he would start with arms stretched apart and then gradually bring them together until they met. His arms would converge as they point to more distant objects.

How could his arms converge if a sighted helper did not guide him? Of course, in the case of smaller objects such as household furniture, tables, and chairs, blind subjects can form a first-hand spatial experience, but, crucially, not a visual one.

In such an experience, nothing is intrinsically visual. Once again, our interpretation of empirical evidence is biased by implicit assumptions.[77] Shape perception is neither visual nor tactile. Sometimes, we access aspects of the world that are neither visual nor tactile. They are supermodal insofar as they are aspects shared by different modalities. Other kinds of supermodal content can be envisaged across distinct sensory modalities—e.g., sequence, rhythm, patterns, syntactical structures, combinatorial rules, and so forth.

It is often held that visual perception is located in the so-called "visual areas." Of course, such areas are called "visual" only because they are used by sighted subjects to access visual objects. They are not *intrinsically* visual. As a result, many scientists implicitly or explicitly think that such areas contain visual representations and that, when such areas are active (as can be ascertained through fMRI), a corresponding visual experience must ensue. Since congenitally blind people use the striate cortex to perform shape-and-position related tasks, they are assumed to access the same thing that is generated in a sighted person's mind when they see, i.e., a bunch of neuronal activity in the visual cortex.[78] Sometimes neuroscientists endorse the idea that cortical areas—since they are used for vision in sighted subjects—ought to provide some kind of vestigial vision even in congenitally blind subjects. They are taken to be innately visual. Many neuroscientists, however, do not draw that conclusion (for example, Pietro Pietrini and his group at the University of Pisa). Yet, a popular belief holds that the activation of "visual" cortices is evidence of visual experience, both in the sighted and in the blind.

The above considerations foster a common fallacy. Because the striate cortex is associated with vision in sighted subjects, it is the alleged *visual* cortex in all subjects, blind subjects included. This is a conceptual mistake. If the subject is congenitally blind, the causal link between the striate cortex and optical phenomena has never occurred. The striate cortex in the blind is not a visual cortex. It might have become a visual cortex, but due to lack of sight, it did not. One cannot draw any inference about the existence of mental visual imagery from evidence of neural activity in the striate cortex of a congenitally blind subject. Of course, such questions leave untouched the outstanding empirical evidence of cross-modal plasticity allowing the recruitment of cortical areas devoted to vision in sighted subjects.[79]

The bottom line is that *in a congenitally totally blind person, the striate cortex is not a visual cortex*, even if it is recruited to perform tasks that have cognitive similarity with visual tasks in sighted subjects.

· · ·

Not all hallucinations are related to objects and visual images. People occasionally report hallucinating abstract forms and patterns that do not seem to be part of our world. Are they evidence of pure mental content? For instance, if one has migraine aura, one will experience fortification-like patterns. What are the corresponding physical objects? If one's eyeballs are gently pressed, one will see geometrical patterns. What are they? Here, I do not address simple phosphenes but the more complex geometrical patterns that people report that they hallucinate. What kind of physical object could they be? Finally, if one consumes psychedelic drugs, one will report abstract mandala-like multicolored patterns. To what objects do they correspond? The theory of the spread mind suggests that—in all such cases—the external object one perceives is one's own perceptual system, which is a physical object.

Let us take advantage of the analogy with the kaleidoscope once more. In most cases of geometric hallucinations, one perceives the structure of the kaleidoscope. The kaleidoscope allows multiple causal pathways between the perceiver and one or more objects in the environment. Because of this causal rearrangement, one perceives a different object, which is the phantasmagoric reality one sees through the kaleidoscope. The rearrangement, as we have seen, is as external to one's body as the kaleidoscope is external.

When you peer inside such an optical device, due to the various symmetries and regularities of how mirrors reflect light rays, you see something distinctively hexagon-like. Yet, there is nothing like a hexagon inside the kaleidoscope. However, mirrors, which are external objects too, impose such constraints on the reflections that one sees something hexagon-like in the way in which objects are reflected. The kaleidoscope (like the retinocortical network) determines the organization of what one sees. The kaleidoscope's activity—a result of mirror surfaces—is invisible to the eye. In fact, mirrors are invisible because they always show something else. Thus, when one looks inside a kaleidoscope, one sees the external world disposed in a specific geometry—a common kaleidoscope has a recursive Escher-like hexagonal symmetry. Such a symmetry is not literally the shape of the kaleidoscope but rather the way in which the optical device constrains light rays. In a similar manner, geometrical patterns so common in many

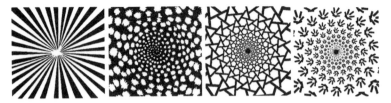

Figure 17. A geometrical rendering of LSD-induced hallucinations (adapted from Bressloff et al., 2001).

different hallucinations are the result of retinocortical topology that determines the geometrical organization of light. Thus, geometric patterns are not concocted images inside one's head. Rather they are composite objects whose structure reflects the structure of the kaleidoscope—namely, the structure of one's retinocortical network.

Consider the example of drug-induced psychedelic hallucinations like those described by Paul C. Bressloff and his group in 2001. It is well known that geometric visual hallucinations have recurring common properties (Figure 17). In 1966, in his book *Mescal and Mechanisms of Hallucinations*, the psychologist Heinrich Klüver grouped these phenomena into four standard groups: 1) gratings, lattices, filigrees, honeycombs, checkerboards, 2) cobwebs, 3) tunnels, and 4) spirals. It is interesting to note that these common features shared by geometrical hallucinations are produced by causes as diverse as pressing one's eyeball or LSD consumption, ranging from direct brain stimulation to dreaming. In 2001, Bressloff suggested that such common geometrical features are determined by "the patterns of connections between retina and striate cortex [...]—the retinocortical map—and of neuronal circuits in V1." They are cautious not to make any explicit reference about *what* one sees in such conditions. Yet Bressloff points out the role of such topological patterns in constraining the geometry of phosphenes. The theory of spread mind suggests that, in all such cases, one sees objects made of past lights reorganized according to the imposed causal geometry of retinocortical networks. The light is in the world. The geometry is in the world too, but it is the physical structure of the sensory organs.

Another related case is the visual aura that frequently precedes an attack of migraine.[80] A visual aura is often described as a kind of fortification because of its geometrical and repetitive organization and geometry. A possible hypothesis, coherent with the theory of spread mind, is that during migraine one experiences local phosphenes—i.e., small hallucinations—whose geometrical organization is determined by the physical connections of neurons in the striate cortex. The migraine aura might be a rare case in which one perceives one's own brain. More precisely, one does not literally see one's own neurons, but the neural topology directly determines the topology of the visual aura. However, and this is crucial, one does not see the topology of neural networks themselves, but one sees the world reorganized by the topology of neural networks as it happens in a kaleidoscope—light and colors from the world, in addition to their geometric organization from the cortex.

· · ·

One more diffuse myth is that, in special conditions, we perceive colors that do not exist. It is known that our phenomenal chromatic space depends on the structure of the chromatic visual system, on the number and kinds of receptors in the retina, on subsequent processing stages, and on cortical arrangements.[81] For instance, a well-known limitation of human color perception is that certain combinations of colors are not accessible. While one can perceive a reddish-yellowish hue—that is, orange—one cannot perceive a bluish-yellow one. Allegedly, this restriction is the by-product of opponent mechanisms in the early stage of color processing.[82] Yet, such constraints do not imply that colors originated inside the visual system; only that our biology constrains the colors human bodies pick out of the environment.

The details of the issue are quite complex. In 1983, in their famous article *On seeing reddish green and yellowish blue*, Hewitt Crane and Thomas Piantanida claimed that it is possible to overcome such constraints and, thanks to special optical contraptions, perceive normally impossible combinations of colors—for instance, a bluish-yellowish hue. Such a finding—recently replicated in updated experimental settings by biophysicists Vincent Billock and Brian Tsou—is neutral with respect to the physical underpinning of color experiences. The new experiments showed that, in the proper circumstances, the human visual system picks up unusual color combinations. In short, in such cases, one does not see *mental* colors, but rather one temporarily picks up extra *physical* colors.[83] If confirmed, they are cases of augmented color vision rather than cases of mental colors.

Surprisingly, such findings have often been interpreted, both in philosophy of mind and in neuroscience, as evidence of the mental nature of colors. This is puzzling. First, because the empirical case in question requires a complex modification of *external* physical conditions either by using sophisticated optical contraptions or by controlling the external visual stimuli. Secondarily, such findings take advantage of changes in environmental conditions and in behavioral patterns. Thus, one perceives forbidden colors because one is able to pick new physical properties out of the environment and not because one's brain concocts new colors. The available empirical evidence is neutral as to which of the two interpretations should be preferred. Consider the original formulation of the experiment by Crane and Piantanida:

> Some dyadic color names (such as reddish green and bluish yellow) describe colors that are not normally realizable. By stabilizing the retinal image of the boundary between a pair of red and green stripes (or a pair of yellow and blue stripes) but not their outer edges, however, the entire region can be perceived simultaneously as both red and green (or yellow and blue).

Crane and Piantanida's setup is an unusual external setting in which the stimulus moves with respect to one's gaze ("stabilizing the retinal image"). In such conditions, a pair of red and green stripes is seen so that one is causally coupled with such opponent hues in a way that would not otherwise be possible. Because of the ingenious stabilization of the stripes with respect to one's gaze, the subject perceives the external world in an unusual way by merging together hues in new ways. *One sees more colors.* Yet such colors are physical combinations that one accesses thanks to the contraption. The case is more akin to a *colorscope* than to a hallucination. Such colors are forbidden and impossible in the sense that usually one is unable to reach them. Of course, the pairs of opponent colors that the human visual system selects—white-black, red-green, and blue-yellow—have no metaphysical relevance. They are the byproduct of a contingent series of events that took place during natural selection. Another animal species might single out different pairs of opponent hues—e.g., white-black, green-blue, and red-yellow—thereby forbidding orange (yellowish-red) and cyan (greenish-blue). Crane and Piantanida's colors are forbidden only with respect to the contingent phylogenetic path the human primate visual system has followed.

As mentioned, a similar experiment—exploiting artificial stabilization of external stimuli—has recently been performed by Billock and Tsou (2001, 2010). They succeeded in repeating the original experiment and caused a "catastrophic failure" of the subjects' visual system. The subjects picked up combinations of external physical phenomena that are normally "forbidden" by the opponency between color channels. In short, they exploited the blur that occurs whenever the visual field is stabilized to keep the presented color stimuli presented in a fixed position on the retina. The boundaries between stripes vanishes and—unlike in standard perception—colors blend. The resulting color mixture is a yellowish blue: that is, a color that is normally not accessible. These results have been criticized because of the limited number of subjects and the subjectivity of verbal reports since "perceived colors are not 'forbidden colors' at all, but rather intermediate colors."[84] If confirmed, such findings would not pose any threat to a strong realist account of colors. They would be cases in which one picks up colors from the environment, albeit unusual ones.

These experiments induced many scholars to draw metaphysical conclusions as to the capability of the brain to concoct novel colors. For instance, Billock, Gleason, and Tsou (2010) ventured to state that the experiment provided an answer to Hume's missing shade of blue. Interestingly, the case can be used to stress a twofold meaning of Hume's question. On the one hand, the question is about whether one can experience a color that one has never encountered in the physical world. On the other hand, the question is about whether one can experience a new color because, by means of unusual circumstances, one has causal commerce with a hue that is usually not accessible. So far, the empirical findings endorse only the second interpretation since all experimental settings are based on circumstances in which the visual system works differently and thus picks up unusual objects. Thus, in the proper circumstances, one would see Hume's shade of blue because one would pick the hue out of the environment. According to the authors themselves:

> Intriguing, two subjects reported that, after the exercise, they could see reddish green and bluish yellow in their imaginations, although this ability did not persist. We can thus answer the question philosopher David Hume posed in 1739: Is it possible to perceive a new color? It *is*—but the striking new colors that we saw were compounds of familiar colors.

The colors one perceives are the result of an actual coupling with the external world. Imagination—far from being arbitrarily produced by the brain—is limited to "compounds of familiar colors." Imagination is a contact with the property one is acquainted with through the failure of the normal operating mode of perceptual systems. Finally, note that Billock, et al. provide a definition of color composition, stating that "the striking new colors … were compounds of familiar colors."

The bottom line is that such colors are not forbidden in any metaphysical sense, but only with respect to the normal limitations of color-opponent circuitry. They are forbidden in the same way in which turning one's head by 180 degrees is a forbidden movement. Of course, such a movement is forbidden by an average skeletal human structure, but, given a proper "catastrophic failure," it might take place!

. . .

In all cases of unusual experience, a physical object lies at the source. Illusions are cases in which one has wrong beliefs about what one ought to perceive. Hallucinations are cases in which one perceives spatiotemporally remote objects in unusual combinations. Another interesting unusual circumstance is offered by cases in which one's body is altered.

By now, it should be clear that if the body is altered, it will single out different objects and properties from the world. Since experience is identical with the objects that take place thanks to the causal coupling with the body, it will follow that when such a body changes, the objects change, too. If one's sensory structure is modified, one's body will single out different physical properties and objects.

A change in one's physical structure will result either in an addition of unusual objects or in the subtraction of usual ones. Consequently, conscious experience is populated with unfamiliar properties or emptied of familiar ones. Sometimes, the subtraction of a property may result in the perception of a different combination of existing ones. It is pitiful that such circumstances have often been interpreted in terms of mental entities rather than in terms of different carvings of the external world.

Consider a visual analogy (Figure 18). A meaningless pattern of black and white squares is filtered by a grid of white squares. The resulting pattern is a recognizable acronym. By means of subtraction, we will be able to single out something otherwise not accessible. No letters have been added to the original pattern, only something has been removed.

If something is removed from the world we perceive, the resulting world is different. Occasionally, one might have the impression that something has been added. The removal of objects or properties can make something else either more poignant or more easily accessible.

Another possible source of confusion is the temporal displacement that can occur when an object persists for a timespan longer than usual. A classic example is offered by cases in which transducers stop working and one's experience is stuck on the last object perceived. Positive afterimages are a familiar example.

By and large, experiences resulting from alteration in the subject's capacities can be categorized as follows.

- Familiar objects or properties become unperceivable (*subtractions*)
- Relatively unusual objects or properties become accessible (*additions*)
- Past objects or properties are perceived continuously (*postponents*)
- Properties or objects that are normally masked by other properties or objects become easily accessible (*filters*)

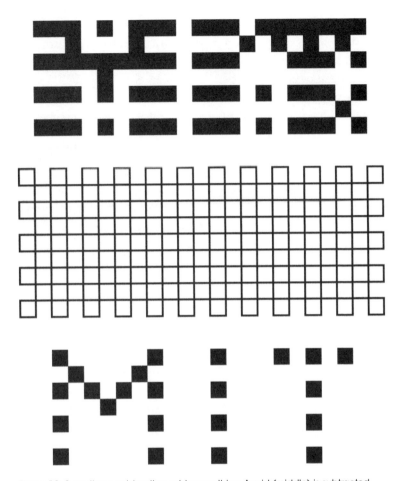

Figure 18. Sometimes, subtraction adds something. A grid (middle) is subtracted from a meaningless pattern (top). The result is a sign (bottom).

The above taxonomy has no pretense of being exhaustive. However, it is useful to consider the manner in which the world can be reshuffled in terms of subtractions/addictions of physical objects and properties rather than in terms of mental entities. Many cases that are usually explained by appealing to additional mental entities can thus be easily revisited in terms of a different carving out of the environment. Not only does one become blind to selected color components. One becomes blind to whole groups of objects and properties—namely, whole groups of objects and properties are removed from the subset of the world that is one's consciousness.

As we will see, the most common cases are represented by subtractions, postponents, and filters. So far, I am not aware of physiological alterations that allow subjects to increase the set of perceivable objects, although certain hallucinogens might indeed work in this way. A lot more frequent is the other case, namely the subtraction of something from one's world because of physiological impairment, sensory stimulation, and cognitive processes.

The actual mechanisms that achieve such deep alterations of perceptual processes are complex. However, with respect to the contrast between appearance and reality, the total disappearance of an object or property is just a shrinking of the subset of external objects that are actual causes of one's internal processes. It is not a change in a mental domain. Remarkable cases of removal of objects or properties are the lilac illusion, motion blindness, and inattentional blindness.

In the case of the lilac chaser illusion, because of complementary adaptation and various mechanisms at various levels from the eyes up to the visual cortex, magenta-ish blobs disappear from one's visual field. A subject is cut out from something that is located in the surroundings. Something is there but it is unable to produce any effect on the subject's body. Something along the causal chain goes amiss. Of course, the external object still produces effects, to some extent, but the causal chain is incomplete as shown by the fact that the subject is unable to use such a perception to check the existence of blobs. At least, when they are invisible.

Analogously, motion-induced blindness is a phenomenon of visual disappearance or perceptual illusions, in which stationary visual stimuli disappear as if erased in front of an observer's eyes when masked by a moving background.

Inattentional blindness is another remarkable case. The famous "invisible gorilla" in the video with the basketball players that Daniel Simons carefully shot in 1999 disappears because a set of expectations prevents it from producing effects in the subject.[85] In these cases, either because of some perceptual or cognitive alteration in the subject's visual mechanism or because of some inherent feature of the object, the external world that one perceives shrinks down to a subset that the gorilla no longer fits into.

• • •

Arguably, the phenomenon that best addresses removal and filtering of properties from one's world is afterimages. An afterimage—or more, generally, an aftereffect—is a change in perceptual experience due to the perception of a previous stimulus. If I stare at a stimulus, my perceptual structures undergo some adaptation and, as a result, I single out different properties. Several different kinds of aftereffects are reported in different sensory modalities.[86]

The theory of spread mind advances a general account of aftereffects—a stimulus changes the causal properties of the body in such a way that the objects and properties that are singled out in the world are different. Explaining such cases in terms of filtering is much more efficient than appealing to mental entities. It is as though the world one usually perceives were filtered through a grid.

The proposed solution is that afterimages are the outcome of a stimulus-induced filtering that singles out different properties in a localized area of the visual field—and thus of the external world. *An afterimage is the perception of a subset of the existing physical properties in a localized region of space.* Since this localized area moves together with the field of view, the phenomenon has often been sketched as a visual image superimposed above the visual field. It is not necessarily the case. In fact, modelling afterimages as the outcome of local filtering devises a way to relocate them in the external world. I separately consider complementary (sometimes called negative) afterimages and positive afterimages.

First, let us consider the most common case—namely, complementary, or negative, afterimages. For the sake of the discussion, consider a classic case of complementary afterimages. You stare for a minute at a red patch and then, when you look at a gray patch, you see a cyanish square (Figure 19). What has happened? The traditional explanation is that you see a mental cyan afterimage since the square is gray and thus the cyan cannot be real. Right? Surprisingly not. Contrary to a widespread tradition, such an explanation is rather naïve since, from a physical perspective, the patch is also cyan, green, blue, red, and many other colors. In fact, gray—like white—is a combination of all color components. It contains all colors in a physical sense.

A different explanation can be put forward. Since you stare at the red patch for a minute, your eyes adapt to red. In other words, you become partially red blind. Because you are now partially red blind in a stimulus-shaped area, you see less red: you stop picking red out of the world. However, the gray patch contains red, green, and blue color components in equal measure. Due to your partial red blindness, you see mostly green and blue. Green and blue together make cyan. That is why you see a cyanish patch. You do not see a mental cyan, you see the physical cyan inside the gray patch. Such a cyan is normally masked by the red component of light. Thus, complementary color afterimages do not require mental colors, but only a different selection of the existing external colors.

Since the area in which adaptation takes place is smaller, the filter affects only a limited area in the visual field (Figure 20). By moving one's gaze, one sees the filter moving with the gaze and thus one can erroneously draw the conclusion that there is a colored mental image hovering around. There is not. Rather, the filtering area moves with the gaze and thus affects different parts of the visual scene.

The difference between the two models is empirically discernible. In fact, according to the spread mind, one only afterimages colors that are physically instantiated by the objects at which one looks. In contrast, according to the traditional view, one afterimages colors that are not physically instantiated—i.e., afterimages are a case of hallucination. In fact, if the colors one afterimages were mental entities, they would depend arbitrarily on one's mental color space. The empirical evidence tells a different story. The color one afterimages is a subset of the physical colors that one stares at afterwards.

Finally, a few words about positive afterimages: they are not the result of adaptation and thus filtering, they are the offshoot of a stimulus so intense that it prevents any further sensation for a while. Positive afterimages are a case in which one is stuck with the last object that one perceived before one was isolated. Often the last object is the one that causes the temporary sensory impairment. A very common case is that of positive afterimages such as those resulting from staring at a bright light. In 1940, the psychologist Kenneth Craik bravely experimented on himself to examine the effects of prolonged observation of the sun. As an effect of such a bold but unwise experiment, he saw the sun for hours after his retina was burned. (Don't do it! He needed six months to recover from the voluntary ordeal of staring at the disk of the sun for two minutes.) Note that he did not see an *image of the sun*, but rather he saw *the sun* for hours.

A causal account of positive afterimages is the following: a stimulus can be so intense that it severely damages the eye. As a result, one ceases to perceive all subsequent objects. One remains stuck to the last event that was responsible for one's neural activity. In this sense, a positive afterimage is as though one were condemned to perceive the last perception one had been able to perceive. Since the last perception was also responsible for the temporary impairment of one's sensorial capabilities, one is frozen to the last causal coupling with the world. One's present stretches until the damaged sensor organs recover their functionality.

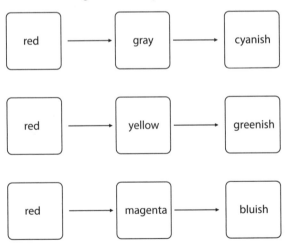

Figure 19. The same stimulus produces different afterimages depending on what the eyes are looking at. Adaptation does not create a new color, but filters the existing colors in the world.

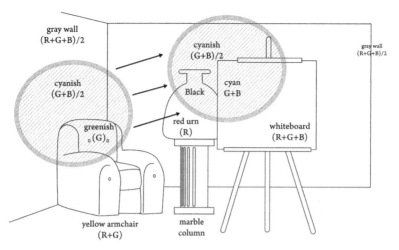

Figure 20. An afterimage is an area in which certain color components are partially filtered off. It is not literally an image though.

In general, Charles Bonnet syndrome, phantom limbs, sensory deprivations, and post-traumatic stress disorders are not different in this respect. Since one or more causal paths are severely impaired, one perceives the last perceived objects. In all these cases, a sudden and severe disruption of one's perceptual capabilities—mostly due to central nervous system damage but not necessarily so, as in the two last cases—sticks one in the present. Subsequently, the subjects keep living the causes that were their last experience before the interruption of the causal continuity with the world. It is as though one's present stretched on forever. In the bestselling young adult novel *Catching Fire*, the second book in the dystopian *Hunger Games* trilogy, a fictional character describes the spreading of a traumatic experience in these terms, "While they try to predict what dishes will be served, I keep seeing the old man's head being blown off." Likewise, in severe post-traumatic experiences, subjects are stuck in a never-ending present.

More traditional positive afterimages can be interpreted in the same manner. A stimulus might be so intense that it shuts down the door of the senses, so to speak. As long as the door remains shut, that stimulus remains in one's experience and in one's present as well.

. . .

Among afterimages, a special place is reserved for a popular myth in the philosophy of mind: the case of so-called supersaturated red. Supersaturated red is presented as a sort of super afterimage, a hue so intense that it cannot exist in the physical world and only exists after looking at it—a purely mental color. Several philosophers have accepted such a case as a concrete evidence.[87] It is not.

Supersaturated red is based on a questionable empirical set up. The unquestioning enthusiasm with which such a doubtful and vague experimental outcome has been received is a testament to the strength of prejudices about the nature of the mind. Supersaturated red seemed to offer empirical support to the autonomy of the mental from the physical and was uncritically welcomed by many philosophers.

Consider Mark Johnston's description of the case from his 2004 paper *The obscure object of hallucination* (italics mine):

> One can come to know what "supersaturated" red is like *only by afterimaging it*. While one is afterimaging it, one could compare how much more saturated it is than the reds exhibited by the reddest of the standard Munsat color chips, there before one on the table. Likewise, a painter might discover in hallucination a strange, alluring color, which he then produces samples of by mixing paints in a novel way.

The key passage is that one can—or is expected to be able to—experience supersaturated red *only* by afterimaging it, "a case which implies that, as a matter of empirical fact, the paradigm red—the reddest of the reds—can *only be presented in delusive experience*." Supersaturated red ought to show that phenomenal experience is independent from the physical. This is the crux of the matter. If it were true, it would imply that supersaturated red is a purely mental experience and thus that colors are a mental stuff. Remarkably, though, supersaturated red is empirically unsound, it is only a philosophical myth which will tumble down easily.

Normal-sighted subjects have never seen supersaturated red, no matter what studies have suggested. It is a popular philosophical myth that has gained momentum because of psychologists' and philosophers' belief in the existence of pure appearances. It has thus been glorified as the perfect battering ram to attack perceptual realism.

.   .   .

Another traditional case of mental images being allegedly different from reality is offered by the by-product of the blind-spot. Because of the exit point of the optical nerve in the retina, the visual field has an area devoid of photoreceptors. As a result, unbeknownst to us, we cannot see a thing in one small area in the visual field.

If the patterns around the blind spot are somewhat regular, we will perceive them as though they were continuous from side to side. The classic stimulus is made of two vertical bars. When the blind spot hides the discontinuity between the two bars, one perceives a continuous bar as though there were a single unified bar. Since the visual system seems to fill the hole with the proper missing pattern, it has been suggested that some filling-in mechanism provides the absent visual detail.[88] The traditional explanation is that, somehow, the brain clones a patch in the visual image and then uses such a patch to fill the hole in the visual field, as though this mental image had to be patched. Explanations that are more refined suggest more complex reconstructive skills, but the most popular explanation remains the same: the alleged mental image is actively filled in with information from the background. Notably, a few scholars have put forward alternative explanations. For example, Dennett has argued that the empty area is simply ignored, and consequently active filling-in is not required (Dennett, 1991).

The theory of spread mind suggests a similar solution. Rather than looking for a mechanism of active reconstruction of a missing portion of the (in)famous mental visual field, one never looks inside the missing whole. The spread mind solution to the blind spot is different from the filling-in model and is more akin to the enactivist approach. A location you cannot look inside does not need to be patched. Thus, the impression of seeing a continuous bar is the result of the lack of difference between a continuous bar and a broken one when they are in that position. One may wonder as to why one should see a hole, if one cannot look where the hole should allegedly be. The absence of information is not information of absence. As Kevin O'Regan noted, if one is not aware of what is going on between one's fingers, why should one be aware of what is going on between photoreceptors? One cannot have any kind of causal intercourse with that part of the world. One's environment is simply not made of those events that correspond to the blind spot. The mind does not spread to them.

. . .

Phantom limbs are one more popular battering ram for the internalist. In such a condition, a patient who has lost a limb reports corporeal sensations from it. The theory of spread mind has no difficulty explaining such cases: phantom limbs are cases of delayed perception. One perceives the limb one once had and, since the corresponding neural area is no longer connected to anything, one keeps being causally coupled with past causes.

Phantom limbs are akin to cases in which the current stimulus is absent due to a blockade (sensory deprivation), to a damage in the causal path (Charles Bonnet syndrome), to a stimulus-induced temporary impairment of sensory organs (positive afterimages), or in its own case, to the removal of the external object.

If phantom limbs do not pose any significant threat to the kind of realism as advocated by the theory of spread mind, what about the even more puzzling cases such as phantom limbs in congenitally limbless patients?[89] Indeed, congenitally phantom limbs pose a serious threat to any strong externalist account insofar as they suggest the existence of a dual nature of physical reality: on the one hand, neural activity and chemical signals, on the other side, body image and somatosensory content. According to Saadah and Melzack's work in 1994, "*the brain generates the experience of the body. Sensory inputs merely modulate that experience; they do not directly cause it.*" The key claim here is that the body image is not identical with the brain but rather generated by unknown innate neural structures in the brain.

To debunk the ontological significance of such cases, I contrast body image and body schema. The first is a set of alleged stored phenomenal experiences that are expected to concoct a limb-like experience. The latter is any hardwired structure that supports the development of sensory motor mapping. The latter does not necessarily contribute to the experience of innate phantom limbs.

Regardless of the theory of spread mind, obvious empirical and theoretical issues challenge the notion of innate images. For one, it is not obvious what selective advantage might accrue from an innate phantom limb. Second, it is not clear why innate brain mechanisms should generate anything akin to a phenomenal experience of the body. Third, the notion suggests that the experience of the body is somewhat different both from the body itself and from the sensory-motor patterns instantiated by the body. Even more puzzling is how phenomenal experience, say, the feeling of having an elbow bent at ninety degrees, should be coded in one's genetic material. Why? What could be the purpose of such information if any healthy individual can easily get it from real limbs? Why should there be a costly mechanism providing such information to those exceedingly rare cases of congenital amelia, i.e., congenital limb absence? Paradoxically, such innate information might be of some interest only in those very patients for which it would be of no practical value since they would be congenitally limbless.

In fact, if you are born with standard limbs, you do not need an innate phenomenal image. You get this image from your limbs. On the other hand, if you are born limbless, having an innate body image is useless.

In contrast, the notion of body schema is evolutionary consistent. Having an innate body schema can help to speed up development and sensory-motor learning. It can be priceless for all species that have little time for childcare—e.g., as in gazelles, deer, and dolphins. It does not require the transmission and the storage of useless phenomenal experiences.

Another worry lingers ahead. If we were born with an innate mental image capable of producing the experience of a limb in congenitally limbless patients, the experience might turn out to be—albeit in a very

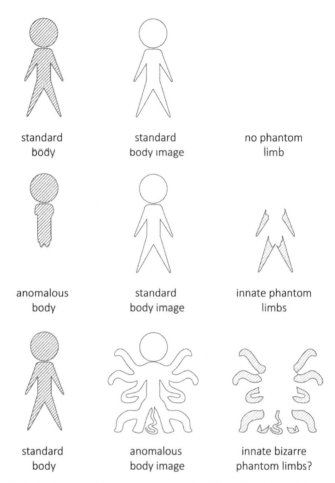

standard
body

standard
body image

no phantom
limb

anomalous
body

standard
body image

innate phantom
limbs

standard
body

anomalous
body image

innate bizarre
phantom limbs?

Figure 21. If congenitally limbless patients have an innate body image, subjects born with limbs might have a bizarre innate body image.

limited number of cases—erroneous. In other words, things might go astray with the innate body image. Sporadically, patients might lament their bizarre, anomalous body image. People might have anomalous innate phantom limbs. This is not the case. Nobody has ever lamented a supernumerary phantom limb or a dragon-like phantom paw or anything like that. To the best of my knowledge, nobody is born with an innate supernumerary phantom limb. Thus, if we had an innate mental body image, I would expect that, in a limited number of cases, subjects with standard bodies might have an utterly bizarre body image and not just a modulation of the standard body image (Figure 21). This is not the case.

More likely, subjects with innate phantom limbs might derive their body image from two easily available sources: their residual sensorimotor input and the observation of other subjects with a standard body structure. It is undeniable that such patients feel an understandable desire to emulate or feel like their peers. As to the former issue, namely the residual sensorimotor input, it is worth acknowledging that no patient has been born so limbless as to be bodiless. All these subjects have some sort of residual somatosensory experience that—at least in principle—can be used to approximate the limb-experience they lack. It would be akin to the well-known experience of flying during dreams. Although it is quite common to experience flying during dreams—either by floating freely in the air or by flapping one's arms—one does not feel anything like what one would feel if one were actually flying, unless one has flown before. Thus, although one might dream of flying, such an experience is not the result of any innate purposeless flying experience mysteriously stored in one's genes. Rather one composes flight, like any object in one's experience, out of other actual experiences—flapping one's arms, floating freely, going forward, moving upward, lying on soft pillows, swimming, and so on.

Phantom limbs are ordinary hallucinations while innate phantom limbs are extraordinary hallucinations. It's very likely that innate phantom limbs have never occurred in the most pristine form as they have been assumed to occur. They are the reshuffling of available sensory-motor experience and strong expectations, organized with the help of an innate body schema.

7.

The Spread Now

You are the music
While the music lasts

—T. S. ELIOT, 1941

ONSCIOUSNESS IS AN intrinsically temporal phenomenon. The word *intrinsic* emphasizes that consciousness is not conceivable outside time, change, and becoming. Consider an obvious analogy, speed, which is an intrinsically temporal phenomenon. In nature, one cannot separate speed from time. One cannot separate a wave from a particle either. One can split the notion of Barack Obama from the notion of being Michelle's husband. One can split the concept, but not the human being. Likewise, no experience occurs outside becoming, change, or time. Speed is change of location. Experience is change of what one is. *Experience is change in what exists, namely becoming.* Change is becoming. Experience requires change. The world requires change too. I argue that time is the unfolding of experience and thus of the world. Time is not an external box inside which things happen; or an additional dimension. I wonder whether the notion of time—as something over and above change, experience, and reality—can be set aside. Traditional time is a useful abstraction—akin to the Galilean object or to Newtonian absolute time—to describe how experience, and thus the world we live in, changes and becomes. However, we never experience time as such. We experience change.

Once more I stress that, as a physicalist, everything must fit into nature. My experience of the present, then, must also fit within nature. The conception of nature must be such as to accommodate a present like the one I experience—namely a *now*.

Earlier, I argued that objects are intrinsically temporal as the physical world we are aware of and we live in. In contrast, the notion of static shapes and timeless objects is suspiciously reminiscent of the timeless ideal platonic world. Both science and analytical philosophy—with their emphasis on language, syntax, logic, and mathematics—are inclined towards a geometric and timeless view of time (Smolin, 2013). Time is conceived of as an additional dimension whose relation with physical objects is extrinsic. Timeless static objects do not play a part in our experience and thus can be set aside. They are not part of our world. The apple I grab from the tree is rotting, the river I step into is flowing, the clouds I look at are dissolving, the apparently immutable rock is slowly eroding, the sun is burning, and the fixed stars are hurtling across space at incredible speed. In our universe, nothing is static. Sadly, the most unchanging things are our prejudices, but even they evolve, albeit slowly. As Walter Benjamin wrote in *The Work of Art in the Age of Mechanical Reproduction* (1935), while the superstructure evolves more slowly than its physical foundations, eventually it evolves too. It is both unfortunate and misleading that—mostly because of the influence from computer science and mechanics—we often refer to the mind in terms of *mental states* as though one's experience were a series of disconnected static frames. In contrast, many authors—most of them with phenomenological tendencies—have stressed the unfolding nature of experience, authors such as Mach, James, Whitehead, Gibson, Husserl, and Merleau-Ponty.

The objective of this chapter is to outline a notion of the present that is consistent with one's experience and is compatible with the physical world. Currently most of the philosophical discussion about time focuses on the tensed vs. tenseless debate.[90] Members of the first group are also known are A-Theorists, or presentists. Those of the second group as B-theorists, or eternalists. The former endorse an irreducible ontological difference of the present. In this regard, Yuval Dolev remarked in 1989 that "the distinction between past, present, and future pertains to our experience and to the way we think and speak." Here I do not choose any of these camps. I address the ontology of the present. The present is not a punctual mathematical abstraction but an extended chunk of reality. The present is a humongous spread object. *We are presents or nows.*

In our experience, everything is there at once and yet everything is the outcome of a change. All we experience are differences in the world. We cannot step out of time, out of change, or out of causality. We never step out of the present. Nor are we ever left in the past. We never step out of the present because we are our present: we cannot step out of ourselves. We are one and the same with the object we call now. Now is an object. Consciousness cannot happen outside of the now because to do so it should happen outside itself. This necessity is the symptom of a fundamental identity. The present is not just a temporal notion, but rather it is a multifaceted notion that refers to the unfolding bubbling surface of nature. The present—or *now* as I prefer to call it—is neither a point on a line, nor a place. I will use the word *now* as a countable noun. I will use sentences like "a now," "many nows," and "each now." The *now* is—for lack of better words—a chunk of reality. *I am not, now—I am a now.* Such a chunk is singled out by causal boundaries and not by mathematical limits. A now defines an actual causal reference frame—namely, a causal network of actual processes that, by bringing an object into existence, fix a reference frame. The now spreads across multiple objects. I will argue that there is no such thing as a unified absolute now, rather there are multiple nows, each coincident with the occurrence of an object or of a moment of experience. Some of these objects are persons. The resulting notion of the now is relative either to the occurrence of an object or—in a symmetrical way—to the occurrence of an experience.

It might turn out that the notion of time is akin to the notions of meridians or centers of mass, useful abstractions devoid of ontological weight. In fact, many traditional dichotomies are only conceptual: existence vs. causation, time vs. change, and experience vs. objects. They are not ontological gaps. They are cracks in our understanding of nature.

· · ·

Both in science and in everyday life, the most common conception of the present is a moving point on a straight infinite arrow. However popular, such a notion is plagued by several downsides. First, the punctual present accommodates neither our experience nor physical phenomena. Second, unbridgeable gaps open up between separate points on the timeline. Third, an invisible moving-forward present is something additional to the physical world and is neither measurable nor observable. Finally, the punctual present leads to the time-gap problem.

What is the span of the present? Is it really a point? And if it is not, is there any limit? Why should we endorse the notion of a punctual present if nothing in our experience is instantaneous? I suspect that the main reason for entertaining a punctual present lies in the desire to comply with bygone assumptions. Historically, the notion of punctual present received a considerable boost because of both Newton's work on calculus and Hermann Minkowski's space-time diagrams. Although these models differ in many details, their underlying notion of a present is roughly similar—the present is punctual and leaves out past and future. Many scholars conceive of time as though it were an additional geometrical dimension in which the present is a geometrical volumeless location. Yuval Dolev recently observed, in his 2007 book *Time and Realism*, that

> [the notion of a punctual present] seems to have become entrenched. For many, the notion that the present is confined to a volumeless point is self-evident. One reason for this is, I believe, the widespread use, in numerous contexts and connections, of diagrams in which time is presented by a line.

Actually, no matter how influential Newton and Minkowski have been, the punctual present is incompatible both with our experience and with the commonly accepted models in neurosciences.

We have often been bamboozled by movies in which—either by magic or by some other fanciful *deus ex machina*—time freezes in a motionless present. The classic scenario, as depicted, say, in the fantastic *The Hudsucker Proxy,* directed by Joel and Ethan Coen in 1994, shows the hands of the clock tower grinding to a halt. Everything freezes accordingly. Snowflakes stop in midair. Cars stop in the center of the road. Pedestrians are frozen in awkward postures in mid-stride. And so forth. Of course, time cannot be stopped. But if it could, most laypeople believe the universe would remain the same, only frozen—a notion backed up by the belief that existence and causation are set apart. Similarly, many philosophers have argued that time without change is conceivable. Such a thesis is questionable, because experience requires change. I do not deny that one might conceive the logical possibility of time without change. I claim that such a notion of time is beyond the grasp of one's experience and thus empirically useless. For one, as above-mentioned, if time froze, the universe would be empty. All objects with which we are familiar would disappear since they are embedded in causal processes. Light itself would disappear. Everything would be utterly isolated from the rest of the universe.

The universe is held together by causal processes and they require change. Existence and causation are two sides of the same coin. Time is not something over and above the occurrence of events. If time froze, everything would be isolated from everything else. The universe would break down. Connections require time to be bridged. Without time all points would become separate from every other.

A simple Newtonian model of time mandates that time is a line and the present is a point. The point keeps moving forward. Admittedly, from a mathematical perspective, the punctual present has many advantages. One is that time is expressed by a single variable. It is moreover an efficient framework for control theory. Since the seventeenth century, control theory mathematized time and thus, in a sense, explained away change. In principle, one can describe any future development as a set of numbers that are available from the start. All future evolution is contained inside the punctual present. While such a model of present has been largely dismissed by recent developments in physics, its influence is still powerful.[91]

The notion of a punctual present is akin to the notion of the Galilean object. It is a useful but misleading conceptual simplification. It is a concept but it is not a reality. No matter how useful the notion of a punctual present is, it is different from our experience. Our present is thick. The notion of the specious—i.e., false—present has been introduced to explain the difference between the experience one has of the present, which is extended in time, and the alleged punctual nature of the physical present. The usual explanatory strategy we have seen at work in the case of illusions is once more deployed—the mismatch between an entrenched notion and the world we live in is solved by blaming individual experience. Of course, as in the case of illusions, the authority of the elite of savants has dwarfed everyone's immediate contact with the world—i.e., conscious experience. The experienced present, which is the only one, has been downgraded to a specious or false present. Since the present *is held to be* punctual by *science*, if experience shows otherwise, experience must be mistaken.

The theory of spread mind is more democratic. If the present we experience is spread, the present will be spread. Experience must be a part of nature. Only a handful of commentators have considered the possibility that the present can be extended, have duration, be thick.[92] Yet, both empirical evidence and scientific data support a spread present.

In our present, objects, words, whole sentences, pieces of music, sounds, gestures, and actions fit in. They require time to complete. They have a beginning and an end—they are not instantaneous. Consider Father Bourdin's objection to Descartes in the *Seventh Objection to Descartes' Meditations on First Philosophy* (1685):

> I know a man who once, when falling asleep, heard the clock strike four, and counted the strokes as 'one, one, one, one.' It then seemed to him that there was something absurd about this, and he shouted out: 'That clock must be going mad; it has struck one o'clock four times!'

The four strokes fit all together inside one's present. They could not fit within a punctual present. They spread across an extended time span. Likewise, our experience does not fit inside a punctual instant—

objects, words, gestures, sounds. Our *experience spreads in time no less than in space*. In my experience, a punctual *now* is absurd in the same way in which a punctual *here* is. Our experience does not fit in a dimensionless point.

To preserve the alleged punctual nature of the present, the customary move is to appeal to some internal representational buffer in which the extended span of one's experience is instantaneously represented. It is the snapshot view. You assume that the neural underpinnings result in a neural snapshot that fixes a mental state and that, like a frame, fits completely within an instant of time. Such a neural snapshot does not exist; it has never been found. It has neither empirical support, nor theoretical consistence. Neural firings are distributed in time. There is no such thing as an instantaneous state of the brain. Even in the most internalist accounts, the brain needs to sustain several tens of milliseconds of neural firings both to do and to feel anything. The notion of brain state—a notion of an instantaneous snapshot—is empty. The punctual present does not contain any neural firing. It is too thin. It does not even contain a single spike. Spikes require at least 3 to 4 milliseconds to complete. For shorter time spans, the notion of spike is meaningless. In a punctual present, spikes do not fit.

Sometimes, the misleading notion of a brain state is suggested by fMRI images presented as snapshots of brain activities. This too is misleading. First, an fMRI image is not an instantaneous snapshot but an integration of various quantities along several milliseconds. Second, an fMRI is not a snapshot but a statistical model. Third, an fMRI image is not directly related to neural firings but to levels of glucose and other chemicals.[93] Fourth, such images suggest a succession of computer-like discrete states.

The popularity of the notion of mental states, brain states, and computational states reveals a deep and largely publicly-denied fear of change. Change is becoming. Becoming, in the long run, is dangerous. Ominously, the most radical form of change is, of course, death. Better not to move, than to move and to end badly.

People often refer to brain states as though it were possible to take a snapshot of what goes on in the brain, a snapshot that defines what will happen next. It is yet another misleading notion. The notion of brain state derives from a twofold misconception. On the one hand, from control theory where a physical system is modeled as an entity whose current instantaneous state fixes all future behavior. On the other hand, from the notion of a machine state fixed by discrete values evolving according to computational rules, such as Turing machines or Conway's life grids. In fact, both models are wrong. From a temporal perspective, both models are inadequate to address what goes on in biological systems in which there are neither discrete states nor meaningful higher order derivatives.

. . .

The *now* is not punctual. It spreads to accommodate both the world we experience and the physical phenomena science describes. Consider this case. My current *now*[94] is made of a keyboard, a table, my laptop, an ominous cloudy sky, grass, a few trees, a summer breeze, a dog barking in the distance. These entities are not in my *now*. They are my *now*. They are myself too. All these entities take place as part of a whole, which is my current *now*. My *now* is such a whole. It is nothing over and above the occurrence of a whole. I am nothing but such a spread object. *My now is a spatiotemporally spread object, which is the actual cause of some joint effect.* Such an object is my experience. It is my world; it is my current now. Experience, appearance, existence, and causation refer to a seamless unity that is spread to encompass an actual chunk of reality. Call such an extended chunk of reality the spread now. I am such a now. Now I am. I am a now.

Such a model of *now* does not require the same universal all-encompassing *now* as Newtonian physics does. Each causal process has its own now and thus multiple nows occur, each one relative to a different object. In the theory of spread mind, the notion of object is that of a cause that takes place in time. Such an object is a spatiotemporally composite object. In turn, the now is not something over and above the object. *The now is the object.* Moreover, occasionally such objects partake of a set of objects that are nobody's experience. *A now, an object, and experience are three different ways to conceptualize the same reality.*

As with everything else—e.g., perception, experience, and causation—each now is spread out to be identical with everything it includes. Consistently, every now is nothing over and above what happens inside it. A now is not the inside which a clock ticks. A now is the clock. *A now is a composite object fleshed out by causal relations.* To recap, the notion of now does not add anything to the already outlined framework of spread objects, actual causes, and spread minds—the three notions are only different ways to address the same underlying occurrence. *Our experience does not take place now. Our experience is a now.*

The now—as with any other object—is defined by causal processes carving reality at its joints. Every object, as well as every experience, has a different duration, and multiple temporal spans coexist inside the *same* now. As a proof, consider a moment in one's experience. I experience an apple, a tree, a cloud on the horizon, a flash of light, a mountain, a voice speaking to me, a tune playing in the background, the moon, the sun in the sky, and some faintly shining constellation. Each of these physical events has a different length and a different temporal location. At any one time, many events which are temporally located in different positions coexist seamlessly. The time span of each now is variable and determined by causal circumstances. Each now is singled out by an event. *Joint causation singles out nows as well as objects,* the two being the same. The now is the part of reality that is identical with a composite actual cause.

Consider again the beloved red apple, this time located on your table. It is part of your present. Yet, the very apple that is part of your present has just reflected some light rays towards your retina. The time delay is negligible but finite. Eventually, you raise your gaze upward. As it happens, you look at a faraway star in the sky—say, Betelgeuse, which is 642 light years far away. Consequently, the star, too, is part of your present. The star partakes of you, no matter the distance between you and the celestial body, as much as the apple on the table does. The star is a component of your now. In the case of the star, what is the time span of the process that brings your present into existence? 642 years. Although some commentators find it difficult to include the perception of stars among the facts of perception, we perceive them just as we perceive apples. Putting stars in quarantine will not sanitize our model of time and perception. On this issue, Dolev comments that: "There is an exception to the claim that only what is present is perceptually accessible: very distant objects such as galaxies. [...]." I do not agree. Stars are part of one's present. I do not see why very distant objects should be an exception. Stars and galaxies might be an exception for our prejudices but not for nature. With respect to their distance from us, buildings, mountains, clouds, the moon, planets, comets, stars, and galaxies partake in a continuum. For von Uexküll's astronomer, celestial bodies are the everyday environment. This is the lively description of an astronomer's life he provides in *A Stroll Through the Worlds of Animals and Men*:

> High on his tower, as far as possible from the earth, sits a human being. He has so transformed his eyes, with the aid of gigantic optical instruments, that they have become fit to penetrate the universe up to its most distant stars. In his Umwelt, suns and planets circle in festive procession. Fleet-footed light takes millions of years to travel through his Umwelt space. And yet this whole Umwelt is only a tiny sector of nature, tailored to the faculties of a human subject.

Astronomers consider stars a more common part of their lives than standard everyday objects. Furthermore, only accidental circumstances shaped our world so that most objects are either close—like apples, buildings, and mountains—or abysmally far away—like stars, constellations, and galaxies—thereby suggesting a difference between the two groups. There is no difference. We see a star just like we see the sun or a distant star. It is not far-fetched to envisage a scenario in which human beings earn a living in a world in which all the distances are uniformly represented. In such a scenario the lack of any difference would be manifest.

Since the *now* is temporally spread to encompass the objects constituting one's experience, memory is the perception of—and thus identity with—the past. Such a past, though, is still present. Ayer wonders whether the perception of stars gives some credibility to the idea that one might be in direct contact with the past. In 1965, in *Mind*, Frank

Ebersole addressed this issue in a paper meaningfully entitled *How Philosophers See Stars*:

> Ayer says things of this sort in order to lend a kind of scientific respectability to the naive realist theory of memory. Because we see the past, it is not wild to think we have immediate memory contact with the past. [...] This theory of memory is an effort to allay our philosophical feelings that the past is gone, to counter doubts about the possibility of memory with the blunt assertion that we have in memory direct access to the past. When we wonder at the possibility of memory, we are supposed to get some help by looking at the stars and saying, 'The past is not gone; it is right here.'

Yet we do not need to share Ebersole's skepticism. We can consider at face value the possibility that one can be in direct contact with the past—that memory is perception of the past; indeed, that people can be identical with their own past. Furthermore, the theory backs up a realist view of memory with two key hypotheses. First, one is in direct contact with the past because one's experience is identical with the past. Second, the past, one experiences, is still present insofar as it is a constitutive cause of the present. Such extended present is the spread now.

Suppose that, after some time, you begin daydreaming and you get lost in past memories or dreams. What are you experiencing? When and where is your experience when you recall your memories? You remember an animated chat with your grandfather during a winter evening many years ago. Is such a talk not part of your present now?

Post-traumatic disorders are often associated with intense daydreaming of past intense emotional episodes and there is a corresponding reduction in the attention to current events.[95] A dramatic expression of such a continued presence of traumatic events is reported by Paul Chodoff in 1963, in his works about survivors of the concentration camps, as quoted by Oliver Sacks in his book *Hallucinations*. The traumatic events *are* still there and even Sacks somewhat partakes of them. Of course, survivors' bodies and brains are with Sacks but their present is still stuck in those terrible moments when they were prisoners. One's present is stuck in the past because of the force of the events in the not-so-near past.

The present extends into cases apparently as diverse as memory, dreams, and hallucinations. When we recall, are we not engaged with our past? Does our past not partake of our present? *Is past not still present when we experience it?* Does anything loom between such an event and your experience? No, it does not. As in the case of perception, transparency is a feature of memory and dream. Does anything separate you and your memories and dreams? Of course not. Nothing keeps you and your memories apart. Similarly, nothing separates your memories from the past they refer to. At any moment, one is co-present with one's experience. No phenomenal layer covers one's past.

The theory of spread mind blurs the distinction between present and past. More precisely, it sets the distinction aside. *One experiences the past to the extent that the past is still present. Memory is the delayed, yet ordered, perception of the past. Dreams are reshuffled perceptions of one's past.* Since I experience the past, what is called the past is part of the present. The objects of both memory and dreams are in one's present, though. Since experience is identity with an object and since memory and dreams are forms of perception, their objects partake of one's present. Both memory and dreams are nothing but identity with a past that is still present.

A *now* is not a point on a temporal line (Figure 22). Nor is it a pair of brackets singling out a temporal window. A now is a set of objects co-present to themselves. How? By being causes of a joint effect. If things were not co-present, they could not meet. Reality would break down. The past is present. Otherwise, if it is no more, it will not exist.

The account defended here suggests also that present is the way in which the becoming of reality structures itself. The present is a spoonful of reality. The past gives substance to the present. The past is still part of the present as long as it is something. If the past were not part of the present, not only would it be no more, the past would not be at all. It would be nothing at all.

Language reflects the fact that the past we experience is still part of our present. In English, the tense which expresses events that partake of the present but which contains the speaker is called *present perfect*—the very name suggests that such a past completes the present. Consider the two past tenses in English. The simple past tense is for things closed in a past that has no time continuity with the present. The present perfect tense is for things that have happened earlier but whose time span spreads to the present. Such things partake of the time in which one is located. Everything a person experiences should be expressed in the present perfect tense since it is also part of one's present.

The difference between the two tenses matches also the difference between episodic and semantic memory. In episodic memory, one experiences again an event one has experienced earlier in one's life. In semantic memory, one refers to facts that one knows only through linguistic second-hand descriptions. Episodic memory is a condition in which events are still capable of producing effects in one's present. In a sense, then, they are still part of one's present. Semantic memory deals with descriptions that have no necessary causal connections with current events.

. . .

All reality is expected to fit inside the *now*. Hilary Putnam stated that "All and only things that exist now are real," echoing Borges's words in *The Garden of Forking Paths*, "Centuries and centuries and only in the present do things happen." Along the same lines, Michael Dummett observed that "we can only describe [the world] as it is, i.e., as it is now." I agree with them all, but with the crucial proviso that each now is spread actually, causally, and relatively to an effect. This is a critical step, which is often under-defended. In fact, while it is difficult to disagree, the crux of the statement is the scope and the ontology of the now. For instance, if one adopted a punctual present, Dummett's observation would be hardly defensible since the world we experience does not fit into an instantaneous present. The spread now has no such limitations. The present spreads and stretches as much as it needs to. It encompasses everything we experience

Our experience backs this up. Consider Emily on a summer night. Emily is looking at the stars in the night sky, hearing thunder in the distance, listening to a friend talking with her, looking at the Portofino bay and, finally, staring at a faraway constellation. Her present includes all these events and yet these events occurred at different times. They took place at different times with respect to that night she was looking at the night sky. Yet, they have something in common. They partake of Emily's *now*. They are all producing a joint effect inside Emily's brain.

As a result, they contribute to the occurrence of a huge spatiotemporally composite object—namely, Emily's *now* on a summer night. They are a huge joint actual cause spread in space and time. In sum, they are a *now*. Any actual joint cause which is spread in space and time is a *now*. Each object is a *now*, too.

Figure 22. The traditional punctual notion of the now (Δ*t*→0).

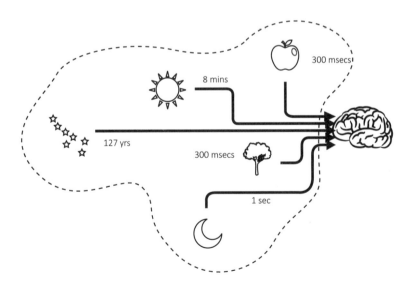

Figure 23. The spread now (Δ*t* can be arbitrarily large).

After all, a spread notion of time is required by internalists too. Aren't neural correlates of one's experience distributed over a time span? Even the most hardcore defender of the brain-identity theory would concede that one's neural underpinnings are extended in both space and time. Then, once we admit that the physical underpinnings are spread in space and time, why should we restrict ourselves spatially to the boundaries of one's skull and temporally to the limits of an instantaneous present?

The traditional notion of past can be set aside as well. Everything is inside a spread present whose structure articulates in various directions. *The difference between past and present is of a practical rather than ontological nature.* Customarily, the part of the spread now that is easier to manipulate is called the present. The other parts are called *past*. Yet, there is no metaphysical difference, just a different level of accessibility. Once again, a practical difference got promoted to the status of a metaphysical gap.

All the events and objects that partake of a whole constitute a *now*. If they were not part of a whole, they would be separate and thus they could not be co-present. In order to be co-present, separate events must bring the same effect into existence.

The *now* is an actual cause—aka a whole, aka a spread object, aka a composite object—whose constituents are at different temporal distances from the final effect. Two events are in the same now if they are germane causes of the same joint effect. Taking place in the same spatiotemporal location is neither necessary nor sufficient to be in the same present. It can help, though.

If the now is carved out in causal terms, any two events belong to the same now when they are in the right causal connection. Two events are not in the same now because they take place inside the same temporal slice. Moreover, there are no temporal slices. Time is the unfolding of a series of nows, causally related.

The proposed causal structure of the now is the one outlined in the case of objects/actual causes—namely joint causation. Such a causal structure has neither a fixed temporal scale nor temporal boundaries. Of course, in practice, close events get causally entangled together more easily. Being close both in space and in time helps to partake of the same causal process and to produce a joint effect. As a result, a good rule of thumb is that what has just happened is often taken to be the present.

According to this account, each co-present event can have a different temporal distance from the joint effect that it causes. Spatiotemporally composite objects such as hallucinations are valid examples. One's now spreads across traditional time and space. Likewise, a perceptual object can be hugely distributed in space as is the case of a constellation made of stars at different distances. The now has multiple and likely heterogeneous temporal lengths (Figure 23). *The present is more like a bowl of spaghetti*—each having its own length—than it is like a point on a line.

It is informative to compare the spread now with Albert Einstein's notion of simultaneity, according to which two or more events are simultaneous only relatively to a given observation.[96] The two notions are very close. According to the layperson's view of time, two events are simultaneous if they take place at the same absolute instant. In contrast, in *Relativity* (1916), Einstein pointed out that there is no such thing as absolute time; the temporal relation between two events is relative to the position and velocity of the observer:

> Events which are simultaneous with reference to the embankment are not simultaneous with respect to the train, and vice versa (relativity of simultaneity). Every reference-body (co-ordinate system) has its own particular time; unless we are told the reference-body to which the statement of time refers, there is no meaning in a statement of the time of an event.

A strong analogy between Einstein's analysis of simultaneity and the spread now is that the notion of being simultaneous or being co-present is relative and not absolute. Here, though, I do not consider frames of reference but meshes of causal processes. With regard to the spread now, the emphasis is not on the frame of reference but on the existence of an actual causal path. However, the two approaches are largely compatible.

Consider Einstein's example of lightning striking two points, A and B. The observer is located at a supposed midpoint M. Simultaneity depends on the contemporary observation of the light at A and B, by an observer at M. In the same text, Einstein objected that:

> [such a] definition would certainly be right, if I only knew that the light by means of which the observer at M perceives the lightning flashes travels along the length A → M with the same velocity as along the length B → M.

The theory of spread mind adopts a different but compatible perspective. Einstein's analysis focuses on velocity to fix the physical properties of phenomena. In contrast, here the focus is on processes carving the unities to which we refer. The theory of spread mind considers everyday causal processes, whatever their speed might be. It also addresses relativistic conditions. In Einstein's account, two events are simultaneous with respect to a third event if they communicate with the third event and if their messages reach the third event *jointly*. Similarly, in the theory of spread mind, two events belong to the same now only if a causal process goes from such events to a third event in such a way that they cause the further event jointly. In both cases, it does not matter whether the velocity along the length A → M is the same as that along the length B → M. What matters is that they get there jointly—i.e., that they are a joint cause.

Einstein's account sets the upper speed limit to the causal processes invoked by the theory of spread mind. Special relativity sets the conditions in which actual communications between a, b, and an observer γ would occur if they took place at the speed of light through straight light rays. Normally, communication, which is a causal process, takes place at a much slower speed than light *in vacuo*. In standard perception, causal processes are embodied in a rigmarole of discontinuous physical hops. For instance, neural processes allow external events to be co-present because they funnel causal processes through dendrites and axons. Such a speed is much lower than the speed of light—give or take, 100 m/s.

The spread *now* is a case of actual simultaneity—that is, it represents a case in which two—or more—objects are unified since they are the constituents of an actual cause. They are co-present insofar as they are brought into existence by the same effect. Very importantly, they are co-present relatively to the joint effect they cause. They are not co-present with respect to other events/observers. Einstein's analysis of simultaneity sets the conditions in which a now—that is an actual cause—occurs. The occurrence of a joint effect is sufficient, though. The theory of spread mind focuses on a world mostly made of slow physical processes such as those involved in standard perception. Actual causal processes substitute the signals envisaged in Einstein's model. The notion of actual cause—that is, the now—is based on that of simultaneity. Light speed signals and relativistic simultaneity are an upper limit. As in Einstein's view, the *now* is relative.

. . .

The idea that the *now* spreads to the past does not contradict empirical evidence. The present is a chunk of reality that encompasses what is usually called the past—it is a huge object. Usually, the discontinuity between present and past is taken for granted. Yet, the metaphysical depth of such a discontinuity has been largely overestimated. In fact—and this is a bit surprising—it is possible to move step by step from our familiar present to our alleged past and back. No absolute ontological gap separates them. The conceptual difference between past and present is akin to the difference between a hill and a mountain—it is the expression of conventional quantitative differences, as is nicely put in the 1995 film *The Englishman Who Went Up a Hill But Came Down a Mountain*, written and directed by Christopher Monger. No ontological difference separates the two. Of course, from a practical perspective, the difference might be relevant. Looking at objects farther away entails stepping into a more remote past.[97]

Consider a critical yet realistic scenario. Call it Major Tom's scenario (Figure 24).[98] Imagine you are chatting with Major Tom through a radio connection. He is on a spaceship travelling towards a distant star. While you chat, he is getting farther and farther away. At the onset of your conversation, your chat with Major Tom is largely unaffected by distance and delay. After some time, though, chatting becomes more and more difficult since the delay becomes greater. The causal distance—i.e., the temporal distance—between you and Major Tom grows greater. Eventually, the delay becomes so great that you can no longer chat with him. The delay is now as great as your lifetime.

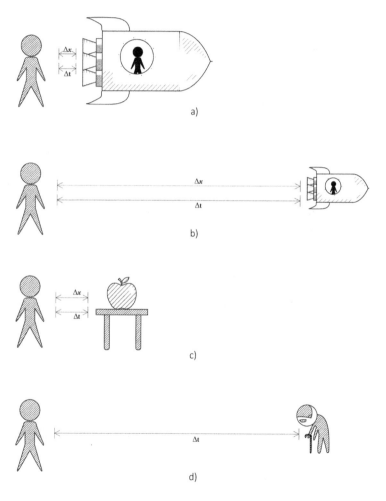

Figure 24. Are Major Tom, the red apple, and one's deceased grandfather in the present or in the past?

Nonetheless, you still listen to Major Tom's words on your radio. In such conditions, it is as though Major Tom's words are coming from the past. Yet, even between the two cases, no ontological gap has been crossed. The two cases are akin to the difference between listening and remembering someone's words.

Past and present are quantitatively different in terms of the time elapsing between phenomena. Yet, they are not ontologically different.

What is the depth of the now? Such a depth varies if one changes the object on which one's attention focuses. While my gaze slides over the visual field, the temporal depth varies—sometimes slightly and sometimes dramatically. By means of a telescope, my eyes peer into the depths of the universe both temporally and spatially. The unaided eye can see objects as remote as the Andromeda galaxy, which is a massive 2.5 million light years away. This fact is often dismissed as a curiosity to amuse kids at the local science museum on rainy Sunday afternoons. Yet it remains an astonishing fact. Read the above sentences again. The unaided eye can *see objects as remote as 2.5 million years ago. One can see the past.* Then, such a past is part of the present.

As mentioned, Ayer stated that when we look at distant stars we perceive the past. Astronomers neither reconstruct the past, nor remember it. Astronomers see the past of the universe, which is still part of the present. They do not see images of faraway stars. They *see* those stars. *One can see the past because the past is still present.* If something is seen, something is part of one's present. Of course, there is a big difference, in practice, between what is called past and what is called present. I see both the part of the spread *now* which is customarily called past, and the part which is customarily called present. Crucially, I can't reach the former, but I can reach the latter.

For instance, in the case of constellations, because of huge spatial distances, one cannot interact with those stars. In everyday life, it is normal to downgrade the distant object as if it were an image or a memory. The lack of reciprocal interaction encourages putting quasars and Major Tom into the past. Yet, the perception of the red apple on the table and the perception of the Andromeda galaxy are not different. Stars are no exception. Major Tom does not recede into the

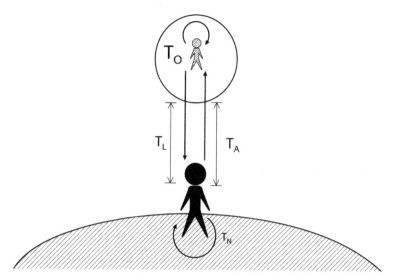

Figure 25. A phenomenon is taken to be present if one can interact with it.

past simply by changing location. You and Tom are linked by causal processes. Sometimes they are comfortably short. Sometimes, they are inconveniently long. No ontological chasm occurs, though. The relevant causal processes and all conceivable possible physical conditions are the same, only more stretched. Quantitative differences do not become metaphysical gaps. Apples and stars stand in the same relation with your body. You see both in the same way and under the same circumstances. The same applies for nearby Major Tom vs. faraway Major Tom.

An astronomical object and the red apple are very different in practice. You can grab an apple, but you cannot grab the faraway star. You chat with Major Tom when he is here but you cannot chat with him when he is very far away. Yet, this is only a parochial difference. A further example will shed some light.

Suppose you want to touch a moving target on a nearby table (Figure 25). You see the target. You move your hand forward, and you touch it. If the target is slow enough, everything works fine. Because of your reaction time, you touch and grab the moving target. Yet, this is not necessarily always the case. Such interaction depends on, at least, four parameters.

Finally, consider placing your target on the moon. Why should a target on the moon not be part of your present? The moon's surface is as much in your present as, say, your room. It is just a bit farther away. Roughly, light takes a second to travel from the moon to the earth. In addition, if you had a powerful enough laser gun, you would be able to hit the moon's surface in one second. For the sake of the example, consider your neural processes as almost instantaneous. Yet, in such a case, you are sure to hit the target only if it moves at such a velocity that it will still be inside the span of the laser beam. Suppose that your laser incinerates everything in a circular area of 1 m radius. You are able to hit objects moving at speeds slower than ½ m/sec. In such conditions, you would be inclined to think that the target is in your present. Faster objects would be perceived as though they were in your past. The example shows that the traditional separation between past and present, although practically useful, is not an unbreakable ontological gap. Both faster and slower objects are in one's present.

Let us move farther away from the earth. Is the sun you see shining in the sky in the past or is it in the present? Of course, as the distance increases, causal interaction—such as grabbing, touching, hitting—becomes increasingly difficult. Yet, no ontological threshold separates what is considered past and what is considered present. No metaphysical gap keeps apart the two extremes—at one end is the faraway star and at the other is someone chatting with you. They are both part of your present.

The bottom line is that no ontological gap separates present from past. Past and present are practical notions based on the conditions of interaction. Something is taken to be in the present if we can interact with it thanks to a favorable time lag.

The past is a part of the present, which is shown by the fact that we experience the past, such as in cases of memory, dreams, hallucination, or perception. If we experience something, it must be in our present.

What about the future? We never perceive the future. Leaving aside cases of premonition or clairvoyance, which I believe are cases unsupported by empirical evidence. When we refer to the future, we refer to anticipations, predictions, long shots, and guesses. Such cases are not any real perception of the future. Rather, they are more or less lucky guesses that one derives from the past.

An interesting outcome of this view of the now is that the present never ends. Since each now is only a now *relative* to a further event, each event can always be the present of some further event. In this regard, *the present never ends.* The things that are happening—here on my terrace—will always be someone else's present, given enough time and distance.

In four years' time, what is going on, here and now, on my terrace, will be part of the present of a beholder who will look at the earth from nearby Alpha Centauri. Each present is relative to a certain observer and to a certain joint effect. In principle, we might imagine the existence of a future observer at the right place and time. Thus, you may imagine that this very moment, which I am experiencing on my terrace (at the end of August 2017), will be present for someone else in the future. It will be part of the *now* of a future observer, not in terms of either reliving the past, or remembering an engram, or rehearsing an image. The *now* singled out by a beholder on Alpha Centauri is not inferior to our own. *The two nows are equally valid.* By and large, an object is still in one's *now* as long as it produces effects. Causal chains can be as long as they like. At least in theory. In practice, when the causal influx is totally expired, the entity no longer exists.

Spatial distance is only one easy way to stretch the time span of such causal chains. Other strategies are feasible. *Nothing prevents us from using other means to stretch causal chains.* For instance, one's neural structures extend the causal influence of events for one's entire lifetime. Thus, through such causal chains, which are propagated by our neurons, we still perceive our past as a part of our present. We call such perceptions memories, or dreams. *We know that a perception is in the past, but we do not experience something as past.* Events in our life produce effects now because of our neural structures. Their causal influence is exerted by means of neural arborizations rather than by means of light rays crossing space and time.

Everything might always be present relatively to some future event. In short, *past is present and everything is now.* Thus, in a sense, the present never ends: the present keeps articulating. The point is that no gap divides past from present. All present is past and all past is present. As T. S. Eliot wrote in *Burnt Norton*, "What might have been and what has been / Point to one end, which is always present." Any moment of being is a now that contains its beginning and its end. *Its end is responsible for the existence of its beginning.* The effect pulls into existence its cause.

Since each now is a chunk of reality singled out by a joint effect, there are multiple nows—and each one is as good as any other. Whenever one experiences something, it exists in the same now that constitutes the subject. The subject and the now are one and the same. Experiencing something means to co-exist with it. More strongly, it means to be identical with it. Thereby experience and co-existence mean taking place inside the same now.

Every time I experience something, something comes into being. That something is my experience, whenever and wherever it is. I am that something. That something is real, although it does not need to be graspable, or made of easily reachable material, or constituted by events in my close proximity, or correspondent to objects other human beings can perceive. A subset of what might be me—and indeed it is me during those circumstances called perception, memory, dream, illusion, hallucination, and so forth—corresponds to the domain we call physical reality. *My now is my world, the kind of thing I am. I am a now. I am a world.*

To sum up. A now is a nexus of events causally entangled and co-present to each other. If this were not the case, events would be miserably and hopelessly isolated. The universe would break down. *Any event or object is part of the now of another event or object, which is the joint effect— there are multiple nows and each is relative to future events or objects.*

·   ·   ·

Coherently with the notion of the spread now, many authors have argued that no intrinsic "cloak of the past" coats one's experience. According to Dolev, "There is no such thing as the pastness of remembered experiences or the futurity of expected experiences." If one dreams of an undefined object—say, a red apple—there is no coating which reveals whether it is an old memory, a recent recollection, or an ongoing perception. In opposition, other authors, such as Merleau-Ponty, have claimed that pastness is a quality instantiated by experience. Experience ought to carry a "sense of the past." I disagree. In one's experience, everything is just what it is. Likewise, in 1951, Furlong comments, "We do not see the pastness of a remembered event in the same immediate way as we see the greenness of a green surface." I would add that we do not see it at all. Time does not deposit a coating of pastness over experience. Of course, one can infer whether something is temporally close by taking advantage of details, events, and sequence of events.

Pastness is not something one experiences phenomenologically. When I experience something by examining it, I do not know whether it is temporally close to what I am doing or not. I can infer its temporal distance only through more or less reliable means. However, *time does not stain experience*. Temporal distance, as usually conceived of, is extrinsic to one's experience. When I look at a star, the star is as much part of my present as anything else. When I recall a friend's face, nothing in my experience tells me whether my friend is an old pal or a recent acquaintance. The point is that no event is intrinsically a past event. I do not see anything that characterizes my experience as a past experience. There are many indirect cues but not pastness. *I infer whether an experience is in the past, but I do not experience pastness.* Perception and memory are transparent. They are moments in which we are in contact with events without any vintage sepia layer over them. We are those events.

In practice, we label experience as memory when we are aware we have already experienced it and we know other experiences took place afterwards. However, such criteria are simply rules of thumb. They are not, strictly speaking, necessary logical conditions characterizing memory. In practice, they induce us to place an event in the past and thus to consider its experience as memory rather than as perception. However, once more, the difference is not ontological. Memory and perception are two extremes of a continuum.

Consider first the lack of interaction. Suppose you are at an interrogation at a police station. You are behind a false mirror that prevents you from interrupting the interview occurring on the other side. At the end of the questioning, the officer tells you that what you have just seen has not happened right now but a few years ago because of some ingenious contraption which uses mirrors capable of slowing light. Would that not be part of your current experience? Would that event not be part of your current now? Would you not have been like an astronomer peering into the depth of the sky? From both a phenomenological and causal angle, no intrinsic aspect of one's experience shows the time of the event you beheld. You have no idea of the time elapsed *based exclusively on experience*. By looking inside one's experience—say, a dream—one has no way to tell when that particular experience is temporally located. One can only guess that something happened earlier because of various clues. However, guessing the temporal location of an experience is akin to guessing the year a building was constructed because of its style. It is a reliable but indirect relation. Sometimes, no cues are available. I experience my children, as they were when they went to kindergarten. Is it a dream or an actual perception? Since I know my kids are enrolled in secondary school, it must be a dream or a memory. Yet, the pastness of my experience is inferred and not experienced. If I had amnesia during the last five years, I could not tell how remote my experience is. Actually, nobody can measure time as something independent from the occurrence of events. Thus, when we say that something happened one minute ago, we mean that, if we had looked at a watch since that thing happened, we would have seen the minute arm only tick once. After all, we *do not see time, but changes in objects*.

Consider the second criterion—comparison among experiences. I experience a face and I remember I saw the face before. Either the same person is in front of me or I am daydreaming that face. It is a rather weak criterion. Still, it is part of what constitutes our sense of the past, which is an inference and not a feeling! Consider what would happen if such a criterion did not hold. Suppose again I had an amnesia that wiped out everything that happened to me in the last five years. Then, I wake up. My experience is stuck in the last moment in which I had been awake before the amnesia. Would that moment not be my present to all effects until I realize I am in a different place and time? We put an event in the past because of various rules of thumb. For instance, the event is unreachable. It must be followed by other perceivable events. We are aware that we have already experienced that event once. Yet, each of these conditions is not the hallmark of an absolute difference. They are useful cues but no more than that: they are practical rules of thumb we apply in order to organize the spread now into a tractable part—the alleged present—and an intractable one—the alleged past. Both parts, however, are the now, the spread now.

Occasionally, we do not have an experience of a direct chain of events between two events x and y. If we wonder about the amount of time between any two events, we cannot measure time as such. We must refer to other events or objects that are chosen conventionally because they are assumed to reliably mirror the flow of time. Here, I use interchangeably the notions of *objects* and *events* since, by objects, I mean a spread object that is causally defined by the occurrence of a causal process. Both events and objects are causes. For instance, we might choose a predictable series of events, such as one's heart rate, the earth's orbit, the day/night cycle, or the oscillation of atomic particles. We assume arbitrarily that such a series of events mirror an invisible clock. However, if we wanted to know whether time elapsed between two events of such a series, what could we measure? If we appealed to another reference series, we would shift the problem of measuring time merely to a different series of events. Alternatively, nothing allows us to know whether time flows between any two such events of a series. *We never measure time: we observe only chains of events.* Since there are plenty of such chains, we are inclined to think that they mirror some invisible abstract temporal unit called time—namely, absolute time, the chain of all chains, pure change with no matter or energy attached. Yet, this ideal series is an empty concept as has been noted countless times as of Newton's work. The classic notion of time is akin to that of real colors—an arbitrary choice of certain phenomena based on parochial and practical criteria. On such an issue, in his popular and short essay *Relativity* (1916), Einstein wrote:

[...] we understand by the "time" of an event the reading (position of the hands) of that one of these clocks which is in the immediate vicinity (in space) of the event. In this manner, a time-value is associated with every event which is essentially capable of observation.

This is a crucial passage. First, time is not over and above the mentioned events, clocks and other stuff. Secondly, the observation fixes a particular frame of reference, while Einstein did not assign any special physical power role to the observer. The link between causes and effects—observations—is key. Third and crucially, the clock has to be "in the immediate vicinity (in space) of the event" so that the occurrence of the clock and that of the event are conveniently placed in the same whereabouts. They have to be co-present. Both clocks and events are causally proximal. Being in the spatiotemporal proximal surroundings makes it more likely that different events are indeed causes of joint effects. Proximity facilitates those events to partake of the same now. Of course, proximity is not a sufficient condition, while joint causal partnership is.

The notion of time as something over and above change is not required except as a very cozy conceptual tool that allows us to know at a glance whether different events are part of a common now. Yet, from an ontological perspective, time is no more substantial than meridians. The arrow of time would not add anything to a collection of events. The notion of an abstract time ticking over and above earthly matters is an invention of linguistic practice.

Traditional linear time is a shorthand to clocks and the like, which are just chains of events that a community chooses for legitimate practical reasons. *We never meet time, we only experience objects.* Objects and their causal interconnectedness are enough. We never see time. We only see clocks. Clocks are objects. Objects are moments of becoming—namely, change. The now—if sufficiently articulated in its causal structure—might let us dispense with the traditional notion of time.

•   •   •

Why do we perceive the flow of time, if we are identical with a now? We are part of nature and we use abstract notions such as time to express the causal unfolding of nature. Yet, time is only a fictitious conventional notion, like meridians and centers of mass. As with other illusions, the notion of the flow of time is the offshoot of beliefs. One measures only the amount of change. Time is akin to the Galilean object—an empty yet useful concept. One experiences useful physical phenomena that are contingently correlated with other familiar phenomena. In such a way one knows, for instance, when the spaghetti is well-cooked by looking at a clock. The heralded subjective flow of time is the perception of chains of objects and events. Sometimes, as is the case with mirrors and lakes, these phenomena diverge from those that have been privileged by legislators and scientists, such as pendulums, mechanical clocks, crystal oscillations, and so forth. In *The Problems of Philosophy*, Bertrand Russell addresses the difference between public and private time:

> With regard to time, our feeling of duration or of the lapse of time is notoriously an unsafe guide as to the time that has elapsed by the clock. Times when we are bored or suffering pain pass slowly, times when we are agreeably occupied pass quickly, and times when we are sleeping pass almost as if they did not exist. Thus, in so far as time is constituted by duration, there is the same necessity for distinguishing a public and a private time as there was in the case of space.

Once again, the received views suggest a conflict between appearance and reality that mirrors a contrast between authority and the individual. The time of the clock is taken to be real while the perceived one is downgraded to mere appearance. Certain phenomena are selected as the alleged real time while others are downgraded to individual subjective times. Contrary to such a tradition, the theory of spread mind sets aside the conflict between appearance and reality as regards time, too. My digital clock, the earth's rotation, my neural processes, and my glycogen levels are equally valid markers of change rather than markers of an alleged absolute time. Of course, they differ with respect to the objects they single out. There is no line to draw between psychological and physical times. They are all measures of change. Of course, some of them are of a bigger practical value.

Consider a familiar experience: for weeks, we have spent our precious time doing the same boring daily chores. Eventually, we look back at those weeks as though time had just evaporated. This notion is nicely embodied by a character in *The Cherry Orchard* by Anton Chekhov. In this play, the decline of an aristocratic family represents the futility of life. The butler who has devoted his whole life to attending the family dies knowing that his existence has been wasted. Like Chekhov's butler, if nothing happened, it would be as though life had never occurred—"Life's gone on, as if I'd never lived." (Act IV).

Contrast the butler's condition with a different case. For a few days, your life is crammed with extraordinary events. Eventually, your recollection of the perceived length of such a period is much greater. It is as though you have lived for a longer time span. If one considers the numbers of events—a measure of change and not of time—the difference is real and it is neither private nor subjective. People have been told that they ought to feel the passing of time, while what they have really felt is the amount of change. Time is an abstraction, but becoming is the reality. Yet, *we do not measure time, we only measure change— i.e., the number of distinguishable spread objects.* Chekhov is right: since change is life, if there is no change, there is no life. Life is becoming, not mere existence.

As in the case of illusions, the community of savants has imposed an alleged time that people ought to perceive. Yet, we never perceive such an alleged objective time, but we perceive other proxy times, that, nevertheless, are as much physical as alleged time.

In Russell's passage, a common mistake happens. He speaks of the "feeling of duration or of the lapse of time" and of the "passing of time." We have neither experience nor measure of time. We do experience or measure the occurrence of some events or objects. We know the earth has rotated once around its axis. We know that, during such a rotation, the short arm of my wristwatch has made two complete turns. I experience the amount of change in one part of my life—what we have called the spread present. However, we have no additional experience of something else that flows. We have no direct experience of the flow of time unless new objects occur.

Conventional time is not objective. It is, however, physical. Moreover, it is not any better than any other experience of change. It is only better in practice. It is the expression of the power of authority imposing a canonical reference to all individuals. Objective time is the society-imposed time. "Objective time" owes its authority to the fact that all clocks have been chosen in such a way as to be consistent among themselves. Biological clocks often wander because they focus on physical and physiological phenomena that may be relevant for biology but that are not important for the human community at large—the level of glucose, the need to sleep, the density of alpha hydroxy acid in one's muscles, and so forth. Yet, objective clocks and one's own biological clocks are on equal footing: they measure the rate of change of physical phenomena, such as atomic oscillations or earth rotations in the former case, and metabolic changes and experiences in the latter one. Objective time is not different from objective temperature. The former is backed up by standard clocks while the latter is supported by thermometers.

Let us bring all the pieces together. On the one hand, the traditional model takes time to be an invisible, puzzling, unmeasurable, causally aloof entity that arches over the whole universe. This entity possesses a punctual now that, like a sort of fleeting cursor, slides over the eons of time. As a result, everything has to be squeezed inside such an instantaneous present with no width. This is inconsistent both with the way in which the brain works and with our experience. Furthermore, it is impossible to determine the temporal location of such a present that, remarkably, plays no role in any physical description of nature. To make matters worse, the notion of a punctual present spawns a series of further puzzles such as the time-gap argument and conflict with our experience that is anything but punctual.

In contrast, the theory of spread mind places change and becoming at the bottom of reality. The now—the present moment—is an object brought together by actual causal processes. The present, which is not a container, spreads to encompass everything we experience. It is in fact identical with everything that one experiences. *The present is an object.* My experience, the world I live in, and the now are three ways to refer to the same moment in nature: a causally singled-out world cell—i.e., a spread object. Such a present, the spread now, neither contradicts any law of physics nor needs any fanciful metaphysical hypothesis. On the contrary, the spread now is consistent with the fact that our experience spreads in time to include past events and is consistent with Einstein's analysis of simultaneity. It is also consistent with quantum mechanics. All moments of our life, as long as we experience them, are still present. Every experience is a perception of something that is present. How could we experience it otherwise, if it were not present? The premises of the time-gap argument are dismissed.

Since the spread now is not an invisible cursor sliding through the eons, when is it located? My bet is that the *now* is always at the end of time. Against the traditional metaphor of time as a river, I wonder whether time is more akin to an ocean. In fact, if time were like a river, where would be the now? We would have no reason to choose a particular point along the course of the river as the current now. Every point would be as good as any other. In contrast, time is like an ocean and the present is its shore. Multiple nows are like separate shores. The ocean, like the river, keeps changing and keeps pushing forward. The shore is always at the end of the sea, though. If you throw a bottle into the sea, the waves will eventually bring it back. The line of the shore is at once stable and changing. As the present, the shoreline is always there. The *now* is not an arbitrary point along a river but the place where all waves reach a conclusion. The *now* is the shore at the end of the ocean of time.

8.

Thou Shalt Have No Other Relations Before Identity

> Nothing can represent a thing
> but that thing itself.
>
> **—EDWIN BISSELL HOLT, 1914**

CCORDING TO A widespread view, the mind represents the world. Every student is taught that representations are numerically distinct from what they represent. For instance, Gilbert Harman expresses this widespread notion by saying that "It is very important to distinguish the properties of a represented object from the properties of a representation of that object. Clearly, these properties can be very different." Indeed, the purpose of representations is having at one's disposal something that, one way or another, acts as a substitute of the object. When the object is not available, its representations are expected to stand vicariously on its behalf.

I do not take issue at the existence of conventional representations, such as traffic lights, voltage levels, chalk marks, hieroglyphs, printed characters, MP3 files, JPGs, and so forth. They are objects, which are used conventionally in the absence of what they are assumed to represent. Such use is parasitical on human beings' arbitrary choices. Neither do I question the existence of functional representations that play a functional and causal role. My point is that, in the physical world, no

217

physical phenomenon can be a mental representation, the only possibility is in being identical with the object.

A separation between the represented object and its representation—so deeply engraved in the philosophical tradition—has been one of the most obnoxious culprits in the scientific and philosophical account of the mind. Poignantly, in 1992 John Searle bragged that the term "representation" is one of the most abused in the history of philosophy. The issue of representation has not been a mere problem. It has been a tragedy.

Once one admits that subjects experience the world, two possibilities are usually taken into account: either that subjects entertain a relation with the world or that subjects internalize something that captures some essential feature of the world. So far, both notions have proved unsatisfactory. On the one hand, it is not obvious what such a relation might be in the physical world. In nature, there are objects, events, and causal processes, but there are no obvious candidates for relations. I have never seen a relation in a physics textbook. Neither have I ever met a relation in my life. On the other hand, the idea that, inside the head, something reproduces the external world without being the external world is very popular but empirically and theoretically untenable. *The cognate notions of relation and of representation have been the contemporary version of the Cartesian pineal gland.* They are fictitious entities introduced to fill a gap between world and experience.

Luckily, between the Scylla of relationalism and the Charybdis of representationalism, a third route, which has seldom been taken into consideration, is viable—i.e., *representations are numerically identical with what they represent.* Of course, I refer to mental representations. Conventional representations—such as signs, letters, computer files— are different from what they represent. Yet, they do what they do because they exploit the existence of humans. Here mental representations have been relocated in the represented objects, which are spread and relative. More precisely, representations are not *in* the objects, they are *identical with* the objects. The machinery of representation—say, neural patterns or electronic levels—must no longer either access or replicate the properties of what they represent. In fact, once the dreaded arguments from hallucination and misperception are set aside, one is back in Shangri-La. For any experience, an actual object occurs. *Experience and objects are numerically and ontologically the same.*

Many scholars of realist and relationalist tendencies have already tried to get rid of the inconvenient representational role of experience by denying that mental representations instantiate the same properties as what they represent. For instance, according to William Fish "when we say that an experience has a reddish phenomenal character, we are not intending to claim that the experience is *literally* reddish, but rather that it has a property that is somehow correlated with what it is like to have the experience's being: *reddish.*" Likewise, Billy Brewer observes that the notion of representational content is problematic and the "direct objects of perception are the persisting mind-independent physical objects we all know and love." Nonetheless, realists do not go as far as considering whether an identity can hold between mental states and objects. Most direct realists claim that between a person—whatever a person is—and a real object a relation must hold. Still, they are hard pressed to find an ontological status both for persons and for relations. Unfortunately, proposing that the relation is simply a matter of "citing the physical object" does not shed any light about the nature of its relation. By the same token, the notion of person is very vague. It is not better than that of the soul.

The theory of spread mind allows us to put forward a notion of identity between representations and what is represented. Such identity is key because it allows us to reformulate the issue of representation without appealing to any dubious bridging relation. Experience is the very object one experiences. In the same spirit, experiences are mental representations. Thus, let us consider seriously the possibility that a representation about red is red! *In the physical world, the only thing that can be a mental representation of a red apple is a red apple.* I am aware, of course, that this proposal runs afoul of most of what is routinely taught about representations. Yet it might be that, once again, a cherished distinction, such as that between the vehicle of representations and its target, was both unnecessary and misleading.

Experiences are mental representations that, in turn, are identical with external objects. A model of representation, which states that a representation is identical with the represented object, avoids the need for a puzzling relation between incongruent entities such as persons or minds and objects or content.

In fact, in the physical world, we never encounter representations. This statement might sound strange given the ubiquitous usage of the term "representation" in everyday life, in cognitive science, in AI, in computer science, and in neuroscience. All such cases are not mental representations but rather entities that play either a conventional or a functional role. They are not representations in themselves but because of the way in which they are used. As a proof of the absence of intrinsic representations in the physical world, consider that we have no way to tell whether a given physical phenomenon is a representation. We cannot tell because representations do not exist. We see a series of ink blobs on a sheet of paper, are they representations or are they just ink spilled by fortuitous accidents? We measure a series of voltage levels inside an electronic circuit. Are they representations of Boolean values? Are they representations of color hues? Are they representations of sounds? Or, deflationarily, are they just voltage levels? How could we test whether something is a representation by means of direct observations and measurements? There is no way. We cannot ascertain whether a physical entity plays the role of a representation. For instance, the same pattern of electronic levels can represent a number, a character, a sound pitch, a pixel, or just nothing. *No experiment can reveal, by means of the knowledge of a thorough analysis of a physical entity, whether such an entity is a representation of another physical phenomenon.*

The whole issue can be revised by relocating representations in the very objects they represent. Consider a familiar cognitive case: while you are walking home, you see a glamorous outdoor commercial sign with a phone number on it. Afterwards, when you are back at home, sitting on your sofa in your living room, you remember the sign and dial the number on a phone. Does this mean that you have had a *number* stored within your brain? Or that your neurons coded those digits?

Not necessarily. Of course, your neural activity extended the causal influence of certain symbols from the sign in the street to your dialing the phone at home. This is only a causal process, though. From the fact that, at the beginning and at the end of the causal chain there are numbers, it does not follow that, along the way, numbers—or representations of numbers—must be physically located inside electrons. It would be a very naïve belief akin to believing that colors and images are stored inside electronic memories or radio waves. Of course, in the brain, neural events propagate the causal influence of external events. This is not different from mechanical gears propagating the causal influence of the steering wheels to the tires.

The same sequence of electronic values in a circuit—identical with the one produced now by my typing on a keyboard—might be the result of random events. Insofar as they have the same values, it is impossible to distinguish the randomly produced series and the typed one. They are physically identical and cannot be distinguished. Here, of course, I am simply restating Searle's Chinese room argument—namely, that information in itself has neither content nor semantics.

A physical phenomenon has the same causal powers irrespective of whether *it* is used as a representation or not—being a representation, in the traditional sense, is epiphenomenal. Thus, representations are physically immaterial—*the notion of epiphenomenal is just the politically correct way to state that something does not exist.* As argued elsewhere in this book, existence and causation are strictly connected. To all extent and purposes, if something is epiphenomenal, it is not physical. It does not have a causal role and it does not partake of the physical world we live in. In short, it does not exist. In fact, there are no pure-epiphenomenal physical phenomena. All physical examples of epiphenomenal properties are always only partially epiphenomenal. For instance, the engravings on gears are epiphenomenal with respect to the engine, but are not epiphenomenal with respect to light rays or to a tactical sensor sliding over them. In contrast, the traditional notions of information, semantics, representational content, and intentional content are entirely epiphenomenal and thus have no place in physics. They are empty notions. The fact that the mind is the only place where epiphenomenal properties have been repeatedly invoked should suffice. If mental representations were real, they should be causally active and hence physical.

Both traditional notions of representation and relation have no place in a physical world where everything is identical with itself. A tree is a tree. A rock is a rock. A human body is a human body. An electronic level inside a CCD is an electronic level inside a CCD. A neural firing pattern is a neural firing pattern. No matter how much one dissects a neural process, no representational content will ever be found. Everything exists because of previous causes but, once it happens, it is what it is. The bird has left the nest. *Within the physical world, only identity rules.* Causes produce new effects and effects define the ontology of their causes. Everything is just what it is. It is not a representation of something else. In the physical world, only one thing can represent another object: the object itself. From the fact that a long chain of events might occur between the object and the body, it does not follow that *something literally transits through the causal chain.* Nor does it follow that a representation is waiting at the end of the causal chain like a golden pot at the end of the rainbow.

We represent the world because we are the world we experience. The representation of a red apple is the red apple. Of course, body and world are different things. Since, though, we are not our body, the difference between body and world does not worry us. After all, we do not experience our brains. We experience the world thanks to our brains. Our brains and bodies allow an object, which is our experience, to take place. The body does not access an otherwise autonomous object. The body brings the object into existence. Our experience is that object.

The long-sought solution of the problem of representation is *identity*. An entity x represents y only if x is y. By the same token, I experience x when I am x. The problem of representation arose because scholars have placed instantiations of representations either inside the mind or inside the body. They assumed that the vehicles of representations ought to be different from what they represented. In the philosophy of mind, in neuroscience, and in the cognitive sciences, the main reason for resisting such a straightforward solution has been the alleged separation between experience and world, between appearance and reality. Yet, once the arguments from misperception have been set aside, identity was enough.

Whenever x represents y, x is identical with y. Using a popular jargon introduced by Susan Hurley in the 1990s, I suggest that vehicle, content, and their references are numerically and ontologically identical. They are one. All relations introduced to tackle separate entities—such as intentionality, content, representation, reference—are realized by the identity relation. What does the brain contain then? The brain contains neural structures that expand the causal influx of external objects. The brain is a causal proxy for external objects. Representations are not inside the brain: they are the very objects they are assumed to represent.

. . .

Presenting identity as the only relation is a move of significant momentum. In philosophy of mind and in cognitive science, many tentative relations have been devised to bridge the gap between mind and world. The claim is that *only identity is supported by the physical world. Identity, though, is enough.*

Explaining the notion of representation by means of that of relation is difficult. Two problems arise. In the physical world, what are relations? As mentioned above, semantics and representations are epiphenomenal, hence they have no causal role—they do not exist. Who has ever seen a relation between two objects, whether familiar objects such as chairs or elementary particles? We observe physical processes and natural forces such as electromagnetism or gravity. We observe physical changes. But relations are not observable and, to the best of my understanding, they are causally epiphenomenal. Relations like semantics have no causal efficacy. Worse, they have no physical counterpart. The notion, expressed clearly by the philosopher Harold Langsam in 2009, that "experiences are relations between material objects and minds" is untenable in a physical world.

A physical entity behaves in the same way regardless of whether a semantic relation (or anything like it) holds between that entity and another one. Syntax drains semantics of all causal relevance. In the physical world, causal overdetermination disposes of semantics and other cognate relations. Relations are ontologically dubious when they are instantiated between physical entities. When relations are posited as a bridge between physical objects and persons, not only they are superfluous, they are preposterous. Semantics, if anything, is identity.

The traditional notion of relation is a conceptual crutch to safeguard a conceptual house of cards from ruinous collapse. Relations are not something that one is going to find in nature. They are a way to describe nature. As such, relations are nice conceptual tools—e.g., such as hierarchical relations among the members of a set—but they are neither the building blocks of nature nor of experience. Using puzzling and epiphenomenal relations does not clarify the nature of the mind. At most, one can point to causal relations.[99]

In sum, inside nature, is there anything that plays the role usually attributed to relations? Yes, there is. It is identity:

A is A

The identity relation looks rather dull in comparison with other relations. Yet, in the physical world, identity is the only relation at our disposal when we want to flesh out atoms, chairs, experience, thoughts, and us. Luckily, the good news is that it is enough.

When I look at a tall silvery poplar bending in the Mediterranean hot summer wind, I do not need ethereal relation linking me to that poplar. Nor do I need an internal mental version of the poplar. Right there, a silvery poplar is bending because of the wind. My visual experience *is* that poplar—and not *of* that poplar.

Often philosophers have been mesmerized by language. Deep metaphysical truths have been drawn from the fact that one can utter sentences like "My experience *of* the poplar." As a matter of fact, the expression "experience *of* something" entails a dualistic picture of the relation between experience, content, and world. Yet, it is only a matter of linguistic practice. We can set "*of*" aside and forget it. Alternatively, one can say "I experience a poplar." I have an experience and the world happens to take place, among the other things, as a poplar. The poplar is me. That is all.

My experience shrunk only to the poplar, I would be the poplar (and not my body). My body is the piece of the physical world that allows that cluster of woods, branches, flowers, and silvery leaves to take place. Why should I continue to give credit to the notion that my experience is *inside* my brain, *inside* my body, *inside* my soul, or *inside* anything distinct from the poplar? My experience, insofar as it is the experience of that particular poplar, is just the poplar. Why should I need anything else?

If my previous efforts to tackle misperception have any merit, then for each representation, we will be able to locate an actual physical object. If this is the case, we will have no reason to resist the notion that a representation is identical with what it represents. The two always co-occur. As we have seen, co-occurrence does not require occurring at the same time. It requires being causally entangled in the right way. What is usually taken to be the vehicle of representation, say a neural pattern, is a causal proxy for the external object. The representation, though, is not the neural vehicle but the external object.

Consider the poplar again. We refer to the poplar-bending-in-the-wind entity in two different ways. Likewise, I refer to the poplar bending in the wind and to its experience/representation as though they were distinct. Such a linguistic practice does not split the poplar ontologically. The poplar remains one thing. The poplar we experience is not made of neurons. It takes place because a certain physical system—made of flesh, neurons, eyes, and other biological machinery—allows it to take place as a cause of a joint effect. The poplar we experience is made of wood, leaves, bark, branches, and water. My experience of the poplar and the bending poplar in the wind are one and the same. A is A.

The proposal is straightforward: my experience represents the poplar by being the poplar. Suppose that we do not know yet what my experience is. As we have seen, the simplest solution is that experience is the poplar itself. Sadly, tradition has contravened such a solution because arguments from misperception have convinced everyone that, in many cases, my experience of the poplar might occur without the poplar. On such grounds, experience could not be the poplar. Yet, it should be clear that empirical evidence has been misinterpreted. The fact is, whenever I experience a poplar, a physical poplar is available, albeit spatially spread. Therefore, we can dismiss the arguments from misperception and embrace the simplest solution—namely, that my experience of the poplar is the poplar. My representation of the poplar is the poplar. Identity rules. Famously, in 1904, William James considered a similar notion in his article *Does "Consciousness" Exist?*:

> The puzzle of how the one identical room can be in two places is at bottom just the puzzle of how one identical point can be on two lines. It can, if it be situated at their intersection [...] Well, the experience is a member of diverse processes that can be followed away from it along entirely different lines [...] In one of these contexts it is your "field of consciousness"; in another, it is "the room in which you sit," and it enters both contexts in its wholeness. [...] What are the two processes, now, into which the room-experience simultaneously enters in this way? One of them is the reader's personal biography, the other is the history of the house of which the room is part.

In comparison with James's notion of pure experience, the theory of spread mind outlines a causal model of the room and outlines how to debunk the arguments from misperception. Furthermore, the theory of spread mind does not need to metaphysically step outside of the physical world. No psycho-physical neutral ontology is required.[100] No neutral monism is invoked either. Physical is enough.

While James was developing his pluralistic ontology, a group of scholars, self-appointed as the "six realists," outlined a similar notion of representation by identity. They were Edwin B. Holt, Walter T. Marvin, William P. Montague, Ralph B. Perry, Walter B. Pitkin, and Edward G. Spaulding.[101] Holt in particular, in his book *The Concept of*

Consciousness (1904), stated with great clarity that identity is the only solution to representation.

> A representation is always partially identical with that which it represents, and completely identical in all those features and respects in which it is a representation. In its more strictly logical aspect, every case of representation is a case of partial or complete identity between two systems.

Thus, a representation is the very thing that is represented. In the same spirit, Holt stated unabashedly that "the concept of representation reduces to that of identity. Nothing can represent anything but that thing itself."[102] The view defended here is similar.

When an event allegedly represents another event, no mysterious relation holds between them. Rather, we hold the wrong belief that one of them must partake of the nature of the other. This is not the case. In a physical world, everything is just what it is. Neural firings are neural firings and the red apple is the red apple. Voltage levels are voltage levels and a keystroke is a keystroke. The alleged representing event does not carry the burden of representation. The so-called vehicle of representation is the effect of the causal process that brings into existence and extends the causal influence of the represented event. The thing that represents an event or an object is the event or the object. A is A. Alleged representations are causal proxies of what they are required to represent.

One might wonder why such an insight has only had extremely limited currency in the relevant philosophical literature.[103] I believe several historical factors have prevented an identity theory of representation from being adequately taken into consideration. First, the usual arguments from misperception have suggested that identity is empirically a nonstarter. Second, the authority of neuroscience suggests that neural activity *must be* the only respectable physical underpinning of the mind. As a result, the vehicles of mental representation are expected to be of neural nature. As a result, James's insight about pure experience and Holt's view about identity were dismissed as empirically-unsupported metaphysical blabber. Consequently, most scientists have only looked for suitable intermediate vehicles of representation inside the brain, from sense data to David Marr's cognitive models, from Wilder Penfield's engrams to Giulio Tononi's integrated information. Lulled into a dualistic dichotomy—mind vs. world—most scientists have dismissed the possibility that representations might be identical with the world they represent. Third, the computational paradigm has conflated the notion of representation with that of information. Most scholars have assumed that information, computation, and representations are inside the head as if the brain were a computer.

In contrast to such a tradition, Jerome J. Valberg noted a few years ago that "if we are open to our experience, all we find is the world."[104] Likewise, the notion of acquaintance, which suggests some sort of intimate connection, is a subset of identity. On this issue, Valberg commented in his 1992 book *The Puzzle of Experience* that:

> Presence (in experience) connotes a kind of direct or immediate availability. An object which is present is right there, available to us. This makes it tempting to view presence as the reciprocal of Russell's idea of acquaintance. That is, an object with which we are (in Russell's sense) acquainted is present in experience; and an object which is present in experience is one with which we are acquainted.

What is such an availability and how is it instantiated physically? Identity is the answer. Without identity, no relation will ever bridge the gap between two ontologically different and numerically distinct entities. Only identity explains our contact with the world and only experience has achieved such an intimacy. *The only way to represent something is to present that something: namely, to be that something.*

·   ·   ·

Once the mind is spread in the world, the appeal to identity between representation and the represented should not come as too big of a surprise. Eventually any account of representation and experience boils down to identity. At the end of the day, whatever experience is, it has to be identical with something. Everything cannot but be identical with what it is. It is a truism that has too often be neglected. In fact, we have been long under the spell of Brentano's intentionality and (in) existent objects—let alone cognate notions such as representation, content, meaning, and aboutness—which promise to connect separate entities without having to pay any ontological price.

In fact, identity is key in many theories of the mind. The theory of spread mind suggests that 1) experience is indeed identical with physical objects, 2) objects are external to the body, and 3) such objects exist because the body allows them to take place. Yet, the appeal to identity has held in the internalist camp too. Most ontological accounts of the mind have been identity theories. Consider two of the most discussed and popular theories of consciousness: sense-data and brain-mind identity theory. Both insist that mental states are identical with certain entities. Both theories are identity theories.

Consider a simplified version of sense-data. Objects trigger sense-data and one perceives something as a result of having those sense-data. Since I do not experience sense-data by means of additional sense-data, to avoid infinite regress, the only explanation is that "having sense-data" amounts to "being those sense-data." To stop infinite regress, one's mind must be identical with the sense-data that one instantiates. To experience is to have sense-data and the relation between a sense-datum and one's experience cannot be anything else but a relation of identity. Otherwise, the problem of how to experience sense-data by means of additional sense-data would arise. Then, sense-data theories are identity theories.

Now consider brain-mind identity theories, which not surprisingly, are yet another example of identity theories. The motivations are roughly the same. If one wonders how and why brain states are perceived, there is infinite regress lurking ahead. Thus, identity with neural activity is invoked. Brain-mind identity theories, with all their problems, have the same explanatory structure of sense-data accounts—i.e., that something is a mental state in virtue of being instantiated by what is taken to be identical with the mind, in this case the brain state. Regardless that nobody has a clue about how the brain can either resemble or be identical with conscious experience, the main appeal of this view is some kind of identity. The key idea is that an experience is identical with the occurrence of a neural process. It is an interesting empirical hypothesis. The relation between neural activity and experience, if there is any, cannot be anything but identity. One's neural activity is supposed to be identical with one's experience. Unfortunately, the hypothesis fails on empirical ground. Brain and mind do not share any property.

The most relevant difference between sense-data theories and brain-identity theory is what is suggested to be identical with experience—neural activity or the external object. Most recent notions such as qualia or information integration complexes are not different in this respect.

Both sense-data theories and brain-identity theories have failed because they have advanced unsuitable candidates. Sense-data theories advance a non-physical candidate while brain-mind theories advance a physical candidate whose properties do not match one's experience. Sense-data theories have failed on metaphysical ground while brain-mind identity theories have failed on empirical ground. Neuroscientists have not yet admitted their defeat in the quest for finding psycho-neural laws. They keep fueling messianic hopes about some forthcoming future revelation.

Identity is the only viable option. Most natural phenomena are explained in terms of identity between phenomena. I explain what temperature is by showing that it is identical with average molecule speed. I explain the origin of species when I show that it is identical with selection, variation, and transmission. And so forth. In the physical world, what is experience identical to? What are mental representations? The theory of spread mind suggests that one's experience is identical with the actual causes of one's neural processes—namely, with the external objects themselves. Thus, one's experience is identical with *the part* of the world that takes place because of the causal commerce with one's body. The brain and the body bring into existence the external relative object that one's experience is identical with.

In conclusion, an identity theory is indispensable because, ultimately, experience is and must be identical with some part of the natural world. Besides, in the natural world, everything is just what it is. In nature, there is no place left for traditional mental representations. Semantic relations do not fit into a purely physical framework since they pretend to do the impossible—namely, *representing something they are not*. The theory of spread mind, on the contrary, endorses identity as the only relation and puts forward a physical candidate for experience and for representations: the relative actual spread object.

Finally, identity is transparency on steroids. Nothing is more transparent than the absence of any occluding medium, because of the identity between representation and represented. Coherently, many scholars have pointed out that no phenomenal layer masks the world we experience; notably the aforementioned Gilbert Harman in 1990.

What could be more transparent than identity? What could better explain the lack of any difference between one's visual experience and the presented object? Identity is the solution. Yet, if one assumed that experience and object—the representation and the represented object—*are* different and numerically distinct, their relation would be hopeless. Identity solves all traditional issues.

•   •   •

The chapter addresses only mental representations. I am perfectly aware that plenty of physical phenomena—either natural or artificial—are used as representations of other objects, events, and facts. I am also aware that states of matter are used as signs, evidence, index, and as measures of other physical events. This is trivial. Yet, such phenomena do not account for the representations that allow us to experience the world. Let me spend a few words to clarify the distinction between different kinds of representations. As I have already done, I distinguish between conventional, functional, and mental representations.[105] Of course, the representations I address belong to this last kind and they are one and the same with one's experiences. I make no claim to put forward any original taxonomy here.

A conventional representation is whatever state of matter one chooses to use to refer to another state of matter. I draw a line on a sheet of paper and declare it represents the Leaning Tower of Pisa. A functional representation is a state of matter that has a functional/causal relation with another piece of matter. I drop a coin in a vending machine and a gear moves so that my favorite beverage drops down. A robot chases a target, and some electronic value is correlated with the target position. Functional representations are beloved by both cognitive scientists and philosophers of mind.[106] Neither conventional nor functional representations endorse an explanation for how content is related to physical entities: at least any physicalist explanation. The conceivable physical underpinnings of both conventional and functional representations are physical entities and, thus, do not have any additional content. Nor have they intentional, teleological, or representational content. In nature, only physical phenomena and physical properties occur.

Both conventional and functional representations owe their importance to external circumstances that do not modify their nature. Like an ace in a deck, they do not change additional properties based on the game we choose to play. A card can stand for anything we like. Likewise, a voltage level in a circuit can stand for music, words, images, and so forth. A ring in a tree is useful to estimating the yearly amount of rain. Of course, one is free to nickname such phenomena as "representations" of other phenomena. However, no elusive "representational content" hides inside such phenomena (whether a sign, an electronic level, or a ring in a tree). Such a functional role does not make the voltage level in any way wordish, soundish, or imagish. Nor does the ring in a tree make the wood in any way rainish. No amount of scrapping, scratching, or probing will ever reveal any content inside them: they have no intrinsic content. Such phenomena are just what they are, whether a sign, an electronic level, or a ring. The same considerations apply for neural activities. Neural activity does not harbor any representational content, no more than any other natural phenomenon can harbor. They are not internal rehearsals of the external world. Neural firings are just neural firings.

Conventional and functional representations—such as a sign or an electronic level or a neural state—do not have any intrinsic content. As stressed, no content can be found by examining the representation itself. No content can be found inside physical stuff. The representational content is simply not there. It cannot be there because it is not a physical thing. Hence, content does not exist. Likewise, a neural activity *does not—and cannot—contain* what one represents or experiences by means of it.

Finally, consider mental representations. We experience the world and that is a fact, a natural fact. Thus, nature must be such that one can experience the world. One experiences objects and other circumstances over and above one's body. Such a statement is not a theoretical assumption. *It describes an empirical fact that the ontology of nature must tackle: namely, the existence of mental representations.* In this context, having an experience of something is having a representation of it. Both cases, of course, boil down to having something as a part of oneself. *We represent the world by being made of the world we experience.*

In this regard, the theory of spread mind is a representationalist theory of the mind—whenever one experiences something, one has a representation of that something. If experiences are not subjective mental states, they always have a target in the physical world. The target is reached by means of identity. One's experience is not an internally secreted ingredient, but it is constituted by the properties of the world we experience. Representation is not achieved by means of some unfathomable and physically-impossible relation, but by identity. So, it should not come as a surprise that, on the one hand, the theory of spread mind is an anti-representationalist theory since the traditional notion of representation—as something that stands for something else—is set aside. On the other hand, it is a representationalist theory as long as one reconceives mental representations in terms of identity. Several formulations are equivalent, such as "S represents *x*," "S experiences *x*," "S has a representation of *x*," "S has an experience of x," "S is x," and "S's experience is x." They are all ways to express that something is part of the subject. The subject is identical with something.

The main reason I stress that the theory of spread mind is a representationalist theory is that it states that no experience is without an object. Every experience is what it is because of the object that it addresses in the world. It addresses an object because that experience is identical with that object. The traditional examples of mental states that allegedly have no representational content—such as depression, moods, and meditative states—are not states without content. Rather, they are experiences that are not easily traceable to obvious everyday objects. Yet, they are mostly experiences of bodily states. In the theory of spread mind, difference in content is always a different object.

Mental representations are physical entities that have the properties of represented objects. Only one thing has such properties, namely the object itself. Thus, mental representations are the very objects or events we experience. The fact that one experiences x because one has a representation of x boils down to the fact that one is made of x. One might object that the brain is a structure capable of instantiating multiple functional representations with the external world. This is obvious. In fact, the goal of the brain is to maintain working functional connections with countless events in the world—avoiding predators, chasing prey, choosing mates, and finding shelter. All the sensorimotor cortices exploit complex functional representations[107]—a scientifically-enthralling and empirically-sound finding. However, no amount of functional representations will ever explain the fact that one experiences the external world rather than one's own brain.

Functional representations can occur without experience. Famous cases are blindsight,[108] cognitive functions performed in absence of consciousness,[109] automatic actions such as driving,[110] cognitive functions performed by the cerebellum and other nuclei,[111] unconscious determinants of actions,[112] and all kinds of functions our nervous system performs without awareness. Copious evidence that functional representations are not sufficient for consciousness is available. Functional representations are structures trying to achieve practical goals.

But function is not feeling as the philosopher Stevan Harnad has repeatedly stressed—feeling is not functioning.[113] What is feeling then? *Feeling is being.* Feelings occur when—because of joint causation—physical conditions determine the occurrence of external objects. *One feels something because one is identical with what one feels,* whether the external object or one's body. Emotions and feelings are somewhat akin to Antonio Damasio's somatic marker—i.e., a certain state of the body that we associate with certain situations as, say, having butterflies in the stomach. What is the relation between functional processes embedded in one's brain and one's mind? The functional structures of the human brain—and likely of many animal brains—provide the necessary causal circumstances to allow the huge network of actual causes that make one's spread mind to exist. They bring external worlds into existence.

What about other organs engaged in functional activities with the external world, such as the immune system or the liver? Do they have a mind of their own? Do they have an independent stream of consciousness? Yes and no. The theory of spread mind dispenses with the need for a conscious flow as something distinct from the world made of objects. Therefore, the liver may well be responsible for the occurrence of some object. For instance, the liver can single out a chemical object such as, say, a certain combination of molecules. Likewise, the immune system can single out a virus or a bacterium. Why should the existence of such objects worry us? As the apple singled out by one's visual system is simply the apple on the table and not a mental entity, so the virus is just yet another physical entity. Of course, one's experience does not need to include all the objects that are causally coupled with one's internal organs. Sometimes, due to purely contingent factors, it might happen, but it does not have to. No deep metaphysical mystery lurks here.

The theory of spread mind suggests a different account for conventional representations too. Consider this case. My grandmother asked my grandfather to plant a holly tree in the garden of their country house. He did it lovingly. Unfortunately, the light conditions and the soil were not up to the task. The holly grew up thin and pitiful. Nevertheless, my grandmother was fond of it. After my grandparents died, my mother was fond of the tree too, for the same reasons. Right now, I am sitting in the garden. I look at the lamentably thin branches and sparse leaves. I would gladly cut it down. I have no personal memory of the day my grandfather planted it. To me, the meaning of the tree is lost, but this is not the case for my mother. However, my mother's fondness is not the outcome of an internal neural activity (although her brain state surely plays a role in her behavior) but rather the offshoot of her parents' actions.

Thus, the tree represents her father's love for her mother, but such a representational content is nowhere inside the holly. Where is the content? Not in the tree. If it were inside the tree, in some way, I should be able to ascertain it after a careful examination. Sadly, I cannot. In my mother's world, though, the tree has a meaning, which is not just a trigger for her memories, but a real meaning connected with the tree. For my mother, no amount of brain stimulation will ever be a substitute for the tree. The meaning of the tree is neither in the tree nor in my mother's brain. The meaning of the tree is one and the same with my grandparents' actions many years ago. The meaning is neither here within the tree nor literally here now. Neither is it inside my mother's brain. The meaning is still located when and where my grandfather planted the tree. The tree is the causal link by means of which those actions are still present to my mother. Their actions are part of my mother's now. My mother wants to preserve the tree and, of course, her brain too because—by means of the two together—she still experiences her parent's love and affection.

Such a link does not work for me, since I was not yet born at the time of the events. Thus, no matter how much time I spend with the holly nor how carefully I examine the tree, I am not going to experience anything by means of it. I might slice the tree in thin foils, scrape its skin, and peer inside its pulp, without ever finding anything. I might do the same, in neurological terms, with my mother's brain. Neither semantic relations nor representational content will show up. The meaning of the tree is a part of my mother's world, which spreads back to my grandparents' actions. Such a world does not overlap with mine, which does not extend so far in the past.

To recap: the holly is the means by which certain events still exert their influence in certain conditions. It is the proxy of a past which is still present. However, neither the tree nor the brain have any content inside. The content is outside the tree. Because of the tree and because of my mother's brain, those past events are still present but only relative to my mother's brain. A conventional representation is such in virtue of the fact that an agent extends the scope of one's own mental representations, which are nothing but the external objects and events. Conventional representations play a role in causally spreading the influx of objects. Causal and functional processes provide the physical support for carving out the actual causes of which experience is made.

. . .

As conventional representations do not have any real content inside them, nor are they related with what they represent by mysterious relations, so information is not a stuff that brings content. Information is just a functional representation of great practical value but it is not a container of anything—neither meaning nor content. What is information then? Information is nothing but a quantitative description of the causal relations between events.

For instance, consider a row of dominos ready to topple on each other (Figure 26, left). If the first falls, they will all fall too in ordered fashion. Do we need anything but tiles and causation to describe what

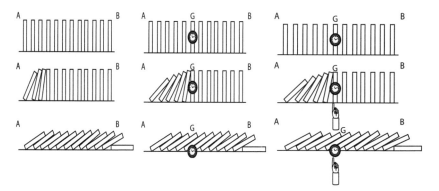

Figure 26. Information is only a description of causal occurrences.

has happened? No, we do not. If A falls, B will eventually fall too. Nothing is shuttled from A to B, only a chain of separate events occurs.

Now consider the second case (Figure 26, center). It is the same as before, only that G is slow. More precisely, the tile (G) has a time delay—when the tile preceding G topples, G topples after a certain time delay. How much? Suppose it can be programmed. Any time span we like, from 1 millisecond to 100 years or more. Is this case any different from the one above? No, it is not. It is just the same, only it may take a longer time. Once again, there is no need to introduce any special stuff carried by the tiles from A to B through G. There are only the tiles toppling on each other.

Last case (Figure 26, right). G not only has a time delay, but it also has a switch that allows an external agent to reset its time delay. Thus, it is possible to make G topple anytime given that the previous tile had toppled already and G is waiting for its time delay to expire. Once again, nothing goes from A to B and nothing waits patiently in G for either some agent to unblock it or the time delay to expire. There are just tiles toppling on each other, a slow tile and the possibility to unclog the causal flow by resetting G's time delay.

But if the above is true, information will not exist! In fact, the above three cases exhaust all that happens in information devices. In particular, the last case is a complete model of both a transistor and a memory cell. Memory cells are nothing but tiles waiting to fall. When we read them, their internal time delay is reset and thus we can see whether they are waiting to fall or not. They are not static containers. They are not boxes. They are not "cells of memory storing a bit of information," they are a channel through which a wave is passing. And the channel has a gate that allows the wave to go through only when one wants to.

If you consider the classic picture of a transistor, which is the basic building block of any computer, and rotate it 90 degrees counterclockwise and compare it side by side with the above dominos, you will immediately see the analogy (Figure 27).

The two cases have exactly the same structure. There is a casual chain from A to B and a gate G that allows the causal chain to propa-

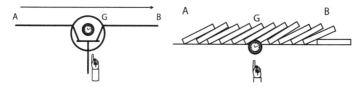

Figure 27. A transistor and a row of domino tiles.

gate. There is no need to add any extra ingredient. There is no information shuttled from A to B. A computer is made only of transistors. No matter how many transistors are there, they do not contain any extra element, no extra juice, no information, no spirit. There is nothing but tiles toppling.

In sum, information does not exist apart as a way to describe what does happen between causally coupled events. This is coherent with the original formulation of information as spun forward by Claude Shannon in 1949. The notion of information is a way to describe causal relations and not a real phenomenon.

As mental representations are not real physical entities, nor is information. In both cases, they are not the kind of entities that can be used to bridge the gap between the brain and the world. They are fictitious entities that we get used to dealing with as though they were concrete entities. Yet, this is not the case. Rocks, apples, cars, bodies, and brains are concrete physical entities. Information and representations are not. The latter are not real. Thus, if we want to explain what is our experience, our consciousness, we must find a physical candidate that can be identical with our experience. This is what neuroscientists have tried with the traditional mind-brain identity theories. Their effort was methodologically correct. Unfortunately, it was empirically doomed to failure since neural activity does not have the properties of our experience. In contrast, here we have considered a mind-object identity theory.

The theory presented here dodges the whole issue of representations and information by focusing on a physical thing—the external object—and revisiting it in terms of temporal and relative existence.

Yet, information is often invoked in relation with memory. Don't we store information inside our computers and does not such information

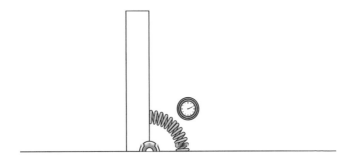

Figure 28. A slightly more sophisticated domino tile.

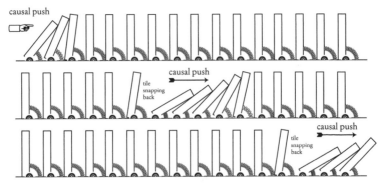

Figure 29. Tiles and information.

carry images, sounds, texts? It should be clear by now that there is no need to posit any intermediate carrier once we set aside the notion of a punctual time. The present is not a point but the whole causal scope of events.

We can thus revisit the notion of information in terms of causal processes. Suppose we have a slightly more sophisticated domino tile (Figure 28). The new tile has a spring with a programmable temporal delay. In practice, whenever the tile falls, the tile snaps back to its original position after a fixed time span. It is not a big change. Rather than having someone putting the tiles back in place, the tiles spring back automatically after a while. Thanks to such a simple contraption, a row of tiles is able to propagate an initial perturbation.

Given that each spring is provided with a little energy, the mechanism is able to propagate indefinitely any initial perturbation for an unlimited amount of time and across an unlimited number of tiles. For the sake of the example, we can represent the row of tiles in a simplified manner. A fallen tile is a black rectangle and an upright tile is a white one. An arrow will show the direction of the causal perturbation. A further simplification is to consider as a fallen tile only the tile that has reached the lowest possible orientation, just before snapping back. The result can be seen in Figure 29. A causal wave propagates from one end to the other. Yet there is no transmission of any physical stuff. What happens at one end determines what happens at the other end. The metaphoric terminology of information transmission is dangerously misleading here. Nothing is transmitted. Things happens as a result of other things. Everything remains where it is. Of course.

What about memory? Doesn't something remain inside a system with memory, as in a brain or computer? Doesn't a trace remain? Of course, something occurs inside a system with memory, but what happens inside a brain is not the carrier of any content. It is rather like a slow propagating causal wave. It is more analogous to a gradually toppling row of dominoes, or tiles. Once again, we can use the example of the falling tiles to outline a model of memory that does not entail any stored stuff.

With a straight line of tile, in principle, the causal perturbation may last as much time as one likes assuming that the line is long enough. The amount of time span is then irrelevant. This is akin to what happens when we stare at faraway stars. It is the sheer length of causal processes that bring in the open the spread nature of the present and the temporal scope of causal phenomena. Yet clearly, in a brain or in a computer, we cannot have a causal chain with a comparable length. Is there a workaround? Yes, and a very easy one, with a sort of causal eddy (Figure 30). Imagine having the following looped configuration of tiles. Once a causal perturbation gets inside it, it will go on forever. Yet, the case will be akin to what happens when we look at a distant star. There is no internal image, no internal movie, no information, no intermediate representation.

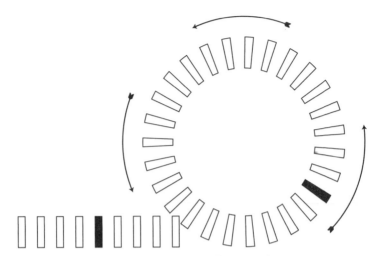

Figure 30. A causal eddy implements memory without any storage.

In this way the past can be indefinitely present since it keeps producing an effect at any future time. Suppose that you have another row of tiles at the other end of the loop, and a gate allows the causal perturbation inside the loop to affect the outside row. By using such simple configurations, we have effectively modelled a physical system where causal perturbations are independent of temporal limitations. A causal perturbation can produce an effect at any later time one likes. In principle, with enough tiles, one can have a system that has a causal presence of an unlimited number of events. It would then be a process-supporting structure in which any event A can cause an event B at any time later. The above shows that, as when we look at a constellation, the cause of what happens can be arbitrarily in the past and yet be present now. No information is required.

·   ·   ·

Leaving aside information, the theory presented here explains the issue of mental representations in terms of identity between our representations and the external world. However, if this is true, how can one represent a chimera—namely something that does not seem to be real? How can the theory of spread mind account for fictitious entities and imaginary worlds? How can one be identical with something that does not exist? How can we account for the ability of the mind to represent imaginary beings? Don't we fall into the risk of postulating the existence of nonexistent objects? I don't think so. As we have seen with hallucinations and dreams, by taking advantage of spatiotemporally composite objects, it can be shown that chimeras are more real than usually assumed. The punchline is that between a chimera and my cat there is a continuum rather than a gap.

In short, to represent something, one does not have either to instantiate some magic relation or to concoct some internal model. Rather, perhaps surprisingly, to represent something one has to let that object happen as part of oneself—i.e., an object is its own representation. For the same reason, representing is pulling something into existence. Then, one cannot represent something that exists independently of one's body. It is not as crazy as it seems.

A useful way to understand the role of brains to create an imaginary-yet-real new world is by considering brains as causal gates. *Brains do not create internal worlds. Brains are causal gates that allow worlds to exist.* They allow the external world to take place in unusual ways that would be otherwise impossible. Of course, this does not imply that brains have any privileged ontological status with respect to the rest of the physical world. They are complex and articulate structures too. In a sense, every physical condition is a gate that allows something else to happen. In turn, everything exists thanks to something else that acts as its own causal gate. Each object exists relatively to other objects. Causation is a gate through which things come into existence. *Brains are causal gates that bring new worlds into existence.* Brains are gates to complex objects, such as states of affairs, relations, patterns, faces, tables, chairs, paintings, cartoons, movies, voices, and so on. Such entities are not meanings *in* the mind—or, worse, *in* the brain. They are actual objects in the physical world. They would not take place if brains did not allow them to happen.

The notion of brains and bodies as gates is particularly challenging in the case of alleged imaginary or mental worlds, such as those created either through imagination or creativity. While imagination is often depicted as a faculty that accesses a pure mental domain, we must admit that all imagined entities exert actual effects in the world. Therefore, they are in the world. They are as real as a rock or as real as a building since they produce effects. They are real objects brought into existence by the appropriate causal circumstances.

A brain is capable of pulling together an otherwise scattered collection of facts. Because of the opportunity offered by a brain, scattered elements take place as a whole. Insofar as something takes place and is the cause of some effect, it is real. Thus, the lizards climbing on the tree next to my window are not different from the fantastic dragons one can only *imagine*. Both dragons and lizards are moments of reality. They both produce physical effects. Hence, they must be real physical entities. We are no more able to create a mental image than we are able to create a lizard out of thin air. Both acts would require godlike powers. However, one's brain allows those parts of reality that compose the lizard to take place in a certain way, which is the lizard one experiences.

Likewise, one allows other scattered events and circumstances, crossed during one's life, to take place as a whole. Such a whole might be a fantastic creature with wings, diamond claws, iron scales, and the like. All these elements, that would not take place as a whole if there were no human brains, are brought together by means of causal intermingling with one's brain. The brain acts like a causal lens focusing a spatiotemporally scattered array of causal processes and objects. The brain brings them together and, by doing so, it allows a dragon to exist. The brain acts like a causal kaleidoscope.

Descartes himself, in his *Meditations*, compares imagination and dreaming to a painter who gathers together different features singled out from the physical world.

> For, as a matter of fact, painters, even when they study with the greatest skill to represent sirens and satyrs by forms the most strange and extraordinary, cannot give them natures which are entirely new, but merely make a certain medley of the members of different animals; or if their imagination is extravagant enough to invent something so novel that nothing similar has ever before been seen, and that then their work represents a thing purely fictitious and abso-

lutely false, it is certain all the same that the colors of which this is composed are necessarily real.

One might raise one or both eyebrows upon hearing that a dragon is real. Real lizards are alive. They climb trees. They can be observed, captured, fed, killed, eaten, and even embalmed. In turn, they breathe, climb, eat, drink, kill, flee, have sex, and reproduce. These are all things that our imagined dragon cannot do. Yet, a sculpture of a lizard would not be able to do it, either. The sculpture can be touched, at least. A painting cannot be touched in the same way a lizard and a sculpture can be touched—the tactile sensation of a lizard and a painted dragon have no resemblance. In this respect, a painted dragon is not so different from an imagined one. To some extent, the *imaginary* dragon could even be better than a painted one since it allows us to touch it in our *imagination*. When one imagines touching a dragon skin, one allows past reptile skins to exert their causal influence and thus to become part of one's present. An imagined chimera is a real object although the causal circumstances allowing it to take place are somewhat unusual. The difference between dragons and lizards is a matter of causal entanglement.

Of course, *practical* differences distinguish one's imaginary-yet-real chimera from a lizard. Such differences are combinations of contingent physical differences in which spatiotemporally composite combinations of events are put together. However, as we have seen, a change in one's perceptual capability might exclude some of these entities. For instance, if one were congenitally blind, one could not experience a painted dragon.

Similarly, many physical objects are private because they cannot be shared. Yet, this hindrance is the effect of contingent factors and not of a metaphysical gap. The rainbow is my favorite example of relative private physical object. More humbly, a hot dog is not so different. One cannot share the very hot dog one is eating since the very act of eating destroys it. Thus, the fact of being private is not exclusive of imagined mental entities. In practice, many physical phenomena are private— rainbows, hot dogs, reflections, visceral states. For instance, although Michelangelo's David at Florence's Galleria dell'Academia is a public object, no two beholders can claim to see it exactly in the same way, due to minute differences in one's perceptual, cognitive, and cultural skills. *Privateness is a matter of degree rather than an ontological divide.* Privateness is a consequence of the relative nature of objects.

Suppose we enter into a museum filled with different animals. Suppose each of us is equipped with a different optical contraption. Suppose, because of these different contraptions, each of us sees the exhibits in a different combination. I see the middle shelves and the bottom shelves while you see the top shelves and the bottom shelves. Each subject reshuffles the museum exhibits (which are an allegory of the world) in a different and private way. However, one does not see imaginary exhibits. Rather, because of our differences, we allow the museum to take place in different ways, thereby making different museums. The fact that different contraptions prevent us from seeing the same combinations does not entail that what we see is not real. Neither does it entail that what we see is mental. *Privateness does not entail an ontological chasm. Privateness entails ownership and relativity.*

By means of painstaking efforts, different people reshuffle their world in original ways, thereby bringing into existence new combinations. After all, this is quite common. Literature is yet another successful way to do it.

Imagination is not so different from handcrafting. One's brain creates actual objects in the world by sheer neural reorganization. This is possible because neurons are embodied in bodies that act as very complex causal gates or proxies for new objects and unknown worlds. The brain is a physical object in a physical world. The brain is no less physical than one's hands. Thus, why should the outcome of the brain be any different from that of the hands?

To recap, imagination is not a process by which one sees a mental world. Rather, *imagination reshuffles the external world by means of causal processes.* Imagination is world making. Imagination is a form of perception. One recombines the actual world in unusual yet actual combinations. They are no less real than other combinations. Furthermore, they are often useful. Once again, *the boundary between real and unreal is drawn at the crossing between useful and useless.* The *real* red apple is the one I can eat. The *imagined* red apple cannot be eaten, yet it can be seen, touched, and smelled! An apple that I could only see appears less real than an edible one. Yet, eating and seeing are two physical processes and the difference is only a parochial offshoot of one's gluttony. Imagination is a physical process that creates new objects. Imagination expands the ontology of the physical world. When we imagine something, our brains act as world makers for that thing to exist.

9.

Look at the Universe and You'll See Yourself

> Things that do not fit into the existing paradigm are hard to think about.
>
> —VINCENT BILLOCK AND BRIAN TSOU, 2010

N WESTERN CULTURE, a popular truism conceives that the mind is a sort of well—or, less poetically, an ontological dustbin or even a digestive system—that first concocts and then stores one's thoughts, experiences, and feelings. The well is illuminated by the spotlight of consciousness. Otherwise, both the psychoanalytic unconscious and the cognitive mind remain in phenomenal darkness. Since such a mind is not as observable as everyday familiar objects, many scholars draw the conclusion that consciousness must be inside something. The notion is akin to the innocent idea that, if something is not visible, it must be hidden, as in a pocket—well, dustbin, or stomach. The pocket, in this case, is the mind.

Yet other alternatives are available. Consciousness might be somewhere else. It might be something we observe every day, only we do not know it is our own mind. Like Edgar Allan Poe's purloined letter, it might be hidden in plain sight. Since, on the one hand, people have assumed that the mind is different from objects and, on the other hand, we only see objects—cars, trees, bodies—a consensus that consciousness is not observable has emerged. Such a consensus, though, leads to endless puzzles. In particular, it has led to a curious conceptual twist: consciousness, which is allegedly invisible in the physical world, is taken to be the phenomenon that allows us to perceive the world. Confusion has followed.

It is a situation that appears hopeless, unless we consider a radically different perspective, as I have tried to do in these pages. I know that such a radical change in our basic assumptions will not be easily accepted because, as Paul Feyerabend warned in 1975, "This is not the usual procedure. The usual procedure is to forget the difficulties, never to talk about them, and to proceed as if the theory were without fault."[114] The prevailing attitude today is to accept the separation between consciousness and the external world. In fact, most scientists and philosophers do not question the received framework—what I call naïve materialism or the standard view. Even enactivists distinguish between sensorimotor contingencies and the external world. No matter the losses, so to speak. The respectable attitude is to proceed as if the separation were an uncontroversial fact.

In the last century, scientists put forward a big effort to move from a metaphysical notion of interiority to the apparently simpler notion of being inside a brain. They rejected the immaterial soul but then took a false step: the brain and the body, with the germane notion of embodiment, became a physical substitute of the soul. Laypeople and scholars share the view that, to know what experience is, we have to look *inside* ourselves. *Inside* what? The body and the brain have become the new soul. Information is the new spirit. Such a direction of enquiry, so far, has led nowhere. There is no internal axis to follow, no downward direction to go in. The universe is in front of us rather than inside us. Even better, the universe—or a part of it—is the thing we are. Nothing hides inside us. The universe is all there is. Ontologically speaking, there is not even an outward direction, since the world is identical with itself. There is no inner world towards which we can turn our enquiry. Speaking of inner vs. outer is not that different from speaking of upward vs. downward about the ground. There is no heaven in the sky or hell downward, as there is no world outside and mind inside. Upward, downward, outside, and inside are not ontological directions but conceptual remains of an anthropocentric notion of nature.

The theory of spread mind expresses a fundamentally alternative view. *One's consciousness is the causally-singled out totality of objects that one experiences. One's consciousness is the spatiotemporally composite objects that exist because a human body allows them to take place.* It is worth stressing that—on the right side of such a definition—no mentalistic notions have been used. Everything is physical and one's consciousness is a subset of the physical world. The particular world that one's consciousness is made up of is carved out by the causal process that, by means of a human body, takes place. Thus, one's mind is physical but, crucially, it is not inside one's body. Neither is it inside one's brain.

Nature is composed of elements scattered in space and time. Every element takes place at a certain time and location. Of course, both space and time are nothing but practical ways to refer to how an object is causally related to all other objects. Any group of these elements composes a new whole whenever a joint effect occurs. At any time, the current ontology of the world is open since it will be fixed only by future contingent effects. Here the word object refers both to simples and to wholes.

Nature is the totality of such objects—e.g., the red apple on the table right now. One's body is also an object. One's experience of an object is the object one experiences. A person is an object, just a different object than one's body—numerically different yet ontologically akin. A now too is an object. Everything that happens is an object. My experience is an object.

Experiences are not additional entities with respect to physical objects. An object is not any different because it partakes in one's life. An apple is always an apple, regardless of whether it is part of one's mind. It does not have any extra phenomenal properties. A mind is nothing over and above the objects it is made of. Some objects are called minds—or persons—because of their role in the history of human beings. *A human being is a set of objects held together by a further object, which is the body. A corpse is an object that no longer brings together the set of objects it did when it was alive. A body is an object that does.*

Compare one last time the standard view—i.e., minds are additional entities with respect to the natural world—with the theory of spread mind. The standard view draws a gloomy picture: we are nothing but bodies walking down crowded streets all encapsulating their own private inner mental world. Such bodies concoct conscious experience inside their heads either like a sparkle of still-burning ember or like powerful spotlights generated by an internal lamp. Each body walks in a world devoid of such light. The world is not only blind and dark, but also mute and deaf since all colors, sounds, lights, feelings outside of these bodies are nothing but mere words. Such a world is a zombie world albeit populated by solipsistic minds imprisoned inside their bodies.

In this view, the first era of the universe was the period when energy was so dense that light could not propagate. Then, after a period of expansion, energy density allowed the first light. Yet, the phenomenal darkness continued long afterwards, in what might be called the unconscious universe, the universe before the birth of the first conscious being. In such a gloomy universe right up until the first sentient beings, there were no colors, no smells, no sounds, and no qualities. Then, fast forwarding to today, a whole stampede of more or less sentient animals and human beings have filled the earth from pole to pole, each of them harboring a special spotlight that has brought the colors, sounds, tastes, and smells of quotidian experience into existence. Each nervous system is ignited within by the conjoined efforts of billions of neurons and synapses transforming their continuous chemical buzzing into the miracle of human experience. As Pascal stated in his cherished *Pensées*, "Man is but a reed, the feeblest thing in nature; but he is a thinking reed." We might be weak but we have booked a central and special place in nature. The sparkle of the mind is ignited in an otherwise dull and brute world. We are special in the universe!

Convincing, reassuring, and flattering, isn't it? Unfortunately, it is also irredeemably false.

In such a picture, the subject is conceived as both numerically different and ontologically separate from the object. Yet, in a physical world, there will be no space for not-physical subjects. Subjects and objects cannot but be two separate concrete physical systems, each in its own space-time location. Between them, relatively minute particles of matter and quanta of energy are exchanged. Thanks to them, one of the two (the subject) concocts in a mysterious way a *replica* or a *representation* of the external object. I use the word *replica* because, once you strip away all the mentalistic jargon, a representation must be a physical entity. The relational option is a nonstarter since there are no relations in physics. Neither replicas nor representations have ever been spotted in the physical world. Only physical entities—namely, objects—populate it. Thus, subject and object face each other as two opposing separate physical systems with their physical features and properties. The subject is identified with the body. Bodies can rearrange their internal structure in some manner but remain separate from their objects. Short of a miracle, why should the subject, which is an object, experience the object? Why should an object experience another object?

The received view is the offshoot of a series of empirically unsupported hypotheses such as the obnoxious appearance vs. reality separation and the notion that consciousness is inside our body. Can't we develop a completely physicalist view of experience and the world? Fortunately, we can. The theory of spread mind is such a theory. ✓

Nature is made by spatiotemporally-spread objects which are causally singled out. Each object exists or takes place relatively to another object. During everyday perception, what I am is a cloud of objects scattered in space-time. I am neither a neural activity nor the interactions between my body and the world. The time span goes from a few hundreds of milliseconds (the red apple) to 2.5 million years (the galaxy of Andromeda). The spatial distance ranges from a few millimeters to hundreds of kilometers on the earth and, again, to millions of light years if one stares at astronomical objects. When one dreams or hallucinates, the time span is as long as one's lifetime. If one dreams or hallucinates a relative deceased many years ago, the episodes are still as much of a part of one's present as when one perceives a red apple on a table. No time gap separates one's experience from the object one experiences, since the two are identical (Figure 31). A span of time keeps apart the object—whenever it happened—and neural activity.

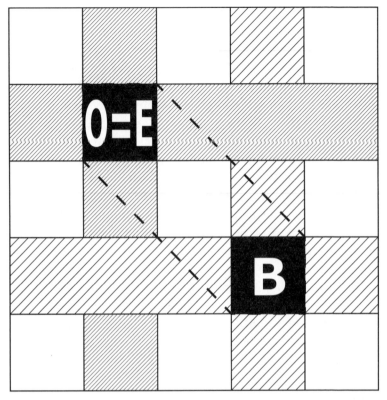

Figure 31.   One's experience of an object is the object itself (E=O). One is not where one's body is. One is where the object one experiences is. Consciousness is physical but not neural.

One's experience is not located where and when one's body takes place, but where and when the object takes place.

Because of the finite speed of physical processes, everything is physically separate from everything else, not only in space but also in time. Thus, the illusion of living in a world of simultaneous events is irredeemably false since the body and the object are necessarily divided both in space and in time. No relation can bridge such a gap. If we—our experience—were inside the body, we would be irredeemably separate from the world. Yet, we—our experience (aka our consciousness)—are the very object and not the body. We do not reach the object thanks to mysterious relations, but because we have always been beyond the gap. We have always been the object on the other side of the object-body chasm.

Because of the speed limits of causal processes, everything we perceive is, necessarily, in a different place and in a different time. Everything, then, is in the past of one's body. Yet, we experience the object and not the body, because we are the object and not the body. We are where and at the time when the object takes place. The illusion of naïve simultaneity tumbles down and reveals the spatiotemporally spread nature of objects and minds; the two being one and the same.

The spread mind outlines a radically different worldview. The world was not dark and colorless before animals and human beings developed. The world was not unlike what we see now. Of course, in the primeval earth, many things were missing, for instance human artifacts. Many animal species were missing and many became extinct later. No complex forms or complex objects were around. Evolution selected beautiful patterns both in terms of animal bodies and in terms of causal structures. Still, the world was not phenomenologically dark. It was not dark at all. In fact, it was filled with physical light, which is the only available light. The very notion of phenomenal experience is a fraud. The world was perhaps simpler than it is now, at least in some respects, but it was otherwise identical. Objects were different and, likely, simpler. Natural selection and brain development have created new forms, new species, new causal structures, and new objects. No abrupt discontinuity marks the development of the first sentient beings: only a continuous and seamless incremental accumulation of objects, animals, and skills. There was no discontinuity making a passage from a phenomenally dark to a phenomenally rich universe. There were objects before us and there are only objects now.

So far, two conflicting worldviews have held center stage in western thought. If one is a hard-core physicalist, the mind is a process inside the brain/body. The body is required to play the role of the soul. If one is ontologically more liberal, the mind is an ethereal entity of some sort, which is peering through the senses at the physical world. Both alternatives, though, are unsatisfactory. The theory of spread mind puts forward a third option. Consciousness is indeed physical all the way down, only, it is not a process inside the body. Consciousness is an object, the huge relative spread object that is made of everything of which we are conscious at a given time. The theory of spread mind explains how two apparently contradictory intuitions are indeed compatible: namely, that we are not our bodies and that we partake of the physical world.

This book stems from a single idea—that one's experience is the external object—and shows how such an idea matches with available empirical evidence. My goal is to tackle both our understanding of nature and our experience of the world. The two must fit together. Furthermore, they must fit with the causal framework that science painstakingly outlines.

Consider an average day in the life of Emily. She wakes at 6:30 a.m. She washes in her bathroom and, in the meantime, her gaze wanders. Absentmindedly, she performs her morning chores in an almost automatic way. Eventually, when she is done, she goes to the gym. While lately it has been a rather cloudy season, quite unexpectedly, today the sun shines gloriously. On the way to the gym, she regrets she has not put on her sunglasses. Consequently, she asks herself a few questions. Has she brought them with her? Nope. Has she lost them along the way? Likely not. Then, where are they? She tries to remember. All of a sudden, she remembers where she last saw them. They were on her bathroom shelf. She has a vivid recollection of seeing them while getting ready. Now, she can almost see them lying on the shelf. Does she see them *in her mind* or does she see them *on her very bathroom shelf*?

The traditional account is more or less the following. During the morning chores, Emily *stores the relevant information somewhere* in her brain without *processing it consciously*. Later, her brain *explores* the *unconscious stored information* until it finds the *information concerning the sunglasses*. Then, on some *mental dashboard*, she is *notified* that the information has been found and, to better inform her, the *information is projected* on a *mental screen* in the form of a *mental image*. Finally, she sees *in her mind an image* of "her sunglasses on her bathroom shelf"—not her sunglasses but *a mental image concocted out of the stuff she has in her memory plus some extra multimedia paint.*

Such an account is extremely complex and requires repeated commuting between mental and physical entities. To my way of thinking, such an account is hopelessly plagued by obscure terms that, no matter how respectable they might appear to a contemporary educated reader, do not offer any true insight about the nature of Emily's experience of recalling her sunglasses.

On the other hand, the theory of spread mind has a different story. While Emily walks to the gym, she acts in accordance with events that have occurred in a larger span. Her brain structure, which is tuned to allow mostly proximal events to exert their effect, changes so as to allow more remote events to exert their influence—until one of those past events is the one that took place when Emily's gaze met the sunglasses on her shelf. As a result, Emily sees her sunglasses on the shelf where she left them. Her glasses are now causally present to her. Since she is aware that her body is no longer in her bathroom, she calls such delayed perception "memory" to distinguish it from cases in which her body can grab what she perceives. In this account, Emily's consciousness is a set of spatiotemporally distributed objects. Her body is the proxy that gives unity to such a set of objects. This set keeps changing because Emily's body keeps changing position and internal structure: for instance, because of neural activity. However, Emily's mind is spread to everything she experiences. Her having an experience is her being identical with such a set of objects. Emily, then, is spread in her spatiotemporal lifetime manifold, while Emily's body is where it is. The body gives to Emily's consciousness the chance to take place, instant after instant. The body, though, is not Emily's consciousness.

The above example shows that there is no need to appeal to mental entities such as stored information to explain how one performs cognitive tasks. Neither does one need mental images stored inside Emily depicting her glasses on her bathroom shelf. The shelf and her glasses are enough to cause Emily's behavior. Her glasses still produce effects in Emily's body *now*. Of course, pressed by closer events, Emily's body requires readjustments in order to allow her glasses to produce effects. Still, what we call a mental process is nothing but a change in the structure of the brain that allows past events to be causally relevant in the present. All cognitive processes are modifications of the causal structure of the world thanks to that astonishing causal bottlenecks our brains and our bodies instantiate.

In the standard view, we appeal to a long list of questionable entities and unfathomable relations whose ontology is the source of endless riddles and whose compatibility with the natural world is murky, let alone that so far no one has a clue how to locate any of them in a physical world—e.g., intentionality, representation, information, character, content, meaning, image, phenomenal experience, and consciousness. In the theory of spread mind, two natural principles—causation and identity—are enough to roll back a series of umpteenth alleged ontological dichotomies that were traditionally considered unbridgeable gaps in the fabric of reality. The list is long: subject vs. object, representation vs. represented, phenomenal vs. physical, appearance and reality, mind vs. world, past vs. present, imaginary vs. real, private vs. public, and hallucination vs. perception. In all such cases, wrong assumptions went unquestioned.

When we look inside our mind, we always reach the world. Even when we recall our remote past, we recall world events, facts, persons, and objects. We never run into a purely mental object. Our experience never stops in the mind or, even more puzzlingly, in the brain. What has been called *introspection* has always rather been a case of *outspection*. Nobody has ever experienced a thought, only what thoughts are about—i.e., objects in the world. The deeper we look *inside* ourselves, the more our experience reaches the world. No interior world looms inside our brain. When we look inside ourselves, we do not see an inner mental world, but the universe itself. We see the universe, because we are the universe we see.

REFERENCES

Albus, A. (2000). *The Art of Arts, Rediscovering Painting*. Berkeley: University of California Press.

Aleman, A., & Larøi, F. (2008). *Hallucinations: The Science of Idiosyncratic Perception*. New York: American Psychological Association.

Aleman, A., van Lee, L., Mantione, M. H., Verkoijen, I. G., & de Haan, E. H. (2001). Visual imagery without visual experience: evidence from congenitally totally blind people. *Neuroreport, 12*(11), 2601–4.

Alexander, S. (1917). Space-Time. *Proceedings of the Aristotelian Society, 18,* 410–418.

Alexander, S. (1920). *Space, Time and Deity*. London: MacMillan.

Allen, P., Larøi, F., McGuire, P. K., & Aleman, A. (2008). The hallucinating brain: A review of structureal and functional neuroimaging studies of hallucinations. *Neuroscience and Behavioral Analysis, 32*(1), 175–191.

Alleysson, D., & Méary, D. (2012). Neurogeometry of color vision. *Journal of Physiology, 106*(5-6), 284–96.

Amedi, A., Merabet, L. B., Bermpohl, F., & Pascual-Leone, A. (2005). The Occipital Cortex in the Blind. Lessons About Plasticity and Vision. *Current Directions in Psychological Science, 14*(6), 306–311.

Anstis, S., & Harris, J. P. (1975). Movement aftereffects contringuent on binocular disparity. *Perception, 3,* 153–168.

Arditi, A., Holtzman, J. D., & Kosslyn, S. M. (1988). Mental imagery and sensory experience in congenital blindness. *Neuropsychologia, 26*(1), 1–12.

Arieti, S. (1974). *Interpretation of Schizophrenia*. New York: Basic Books.

Armstrong, D. (1961). *Perception and the Physical World*. London: Routledge & Kegan Paul.

Arthurs, O. J., & Boniface, S. (2002). How well do we understand the neural origins of the fMRI BOLD signal? *Trends in Neurosciences, 25*(1), 27–31.

Association, A. P. (1975). *A Psychiatric Glossary*. New York: Basic Books.

Attwell, D., & Iadecola, C. (2002). The neural basis of functional brain imaging signals. *Trends in Neurosciences, 25*(12), 621–5.

Aurora, S. K., Welch, K. M. A., & Al-Sayed, F. (2003). The threshold for phosphenes is lower in migraine. *Cephalalgia, 23,* 258–263.

Ayer, A. J. (1940). *The Foundations of Empirical Knowledge*. London: MacMillan.

Ayer, A. J. (1956). *The Problem of Knowledge*. London: MacMillan.

Ayer, A. J. (1967). Has Austin Refuted the Sense-Datum Theory? *Synthese, 17*(2), 117–140.

Barbour, J. B. (1999). *The End of Time: The Next Revolution in Physics*. Oxford: Oxford University Press.

Bartels, A., & Zeki, S. (1998). The theory of multistage integration in the visual brain. *Philosophical Transactions of the Royal Society of London B, 265*, 2327–2332.

Bartels, A., & Zeki, S. (2004). The chronoarchitecture of the human brain—natural viewing conditions reveal a time-based anatomy of the brain. *NeuroImage, 22*, 419–433.

Bartels, A., & Zeki, S. (2005). The Chronoarchitecture of the Cerebral Cortex. *Philosophical Transactions of the Royal Society of London B, 360*, 733–750.

Bednar, J. A., & Miikkulainen, R. (2000). Tilt aftereffects in a self-organizing model of the primary visual cortex. *Neural Computation, 12(7)*, 1721–40.

Benham, C. E. (1894). The Artificial Spectrum Top. *Nature, 51(1313)*, 200.

Bennett, C. M., Wolford, G. L., & Miller, M. B. (2009). The principled control of false positives in neuroimaging. *Social Cognitive and Affective Neuroscience, 4(4)*, 417–22.

Berrios, G. E., & Markova, I. S. (2015). Visual hallucinations: history and context of current research. In D. Collerton, U. P. Mosimann, & E. Perry (Eds.), *The Neuroscience of Visual Hallucinations* (pp. 3–22). New York: Wiley-Blackwell.

Bigelow, J., & Pargetter, R. (2006). Real Work for Aggregates. *Dialectica, 60(4)*, 485–503.

Billock, V. A., Gleason, G. A., & Tsou, B. H. (2001). Perception of Forbidden Colors in Retinally Stabilized Equiluminant Images: An Indication of Softwired Cortical Color Opponency? *Journal of Optical Society of America, 18*, 2398–2403.

Billock, V. A., & Tsou, B. H. (2010). Seeing forbidden colors. *Scientific American, 302(2)*, 58–62.

Billock, V. A., & Tsou, B. H. (2012). Elementary visual hallucinations and their relationships to neural pattern-forming mechanisms. *Psychological Bulletin, 138(4)*, 744–74.

Blakemore, C. B., & Sutton, P. (1969). Size Adaptation: A New Aftereffect. *Science*, (July), 245–247.

Block, N. (1996). Mental Paint and Mental Latex. *Philosophical Issues, 7*(Perception), 19–49.

Block, N. (2005a). Review of Alva Noë's "Action in Perception." *The Journal of Philosophy, 102(5)*, 259–272.

Block, N. (2005b). Two neural correlates of consciousness. *Trends in Cognitive Sciences, 9(2)*, 46–52.

Blom, J. D. (2010). *A Dictionary of Hallucinations*. Dordrecht (Holland): Springer.

Bonneh, Y. S., Cooperman, A., & Sagi, D. (2001). Motion-induced blindness in normal observers. *Nature, 411(6839)*, 798–801.

Borchers, S., Himmelbach, M., Logothetis, N. K., & Karnath, H. O. (2012). Direct electrical stimulation of human cortex—the gold standard for mapping brain functions? *Nature Reviews Neuroscience, 13*, 63–71.

Borges, J. L., & Guerrero, M. (1967). *Imaginary Beings (Manual de Zoologia Fantastica)*. London: Viking Penguin.

Brelén, M. E., Duret, F., Gérard, B., Delbeke, J., & Veraart, C. (2005). Creating a meaningful visual perception in blind volunteers by optic nerve stimulation. *Journal of Neural Engineering, 2(1)*, S22–8.

Bressloff, P. C., Cowan, J. D., Golubitsky, M., Thomas, P. J., & Wiener, M. C. (2001). Geometric visual hallucinations, Euclidean symmetry and the functional architecture of striate cortex. *Philosophical Transactions of the Royal Society of London B, 356(1407)*, 299–330.

Brewer, B. (2006a). Perception and Content. *European Journal of Philosophy*, (1999), 165–181.

Brewer, B. (2006b). Perception and its objects. *Philosophical Studies*, *132*(1), 87–97.

Brewer, B. (2009). How to Account for Illusion. In A. Haddock & F. Macpherson (Eds.), *Disjunctivism: Perception, Action and Knowledge* (pp. 169–180).

Brewer, B. (2011). *Perception and its Objects*. Oxford: Oxford University Press.

Brindley, G. S., & Lewin, W. S. (1968). The sensations produced by electrical stimulation of the visual cortex. *Journal of Physiology*, *196*, 479–493.

Broad, C. D. (1923). *Scientific Thought*. London: Kegan Paul.

Broad, C. D. (1952). Some Elementary Reflexions on Sense-Perception. *The Journal of the Royal Institute of Philosophy*, *27*(100), 3–18.

Brugger, P., Kollias, S. S., Müri, R. M., Crelier, G., Hepp-Reymond, M.-C. C., & Regard, M. (2000). Beyond re-membering: phantom sensations of congenitally absent limbs. *Proceedings of the National Academy of Sciences of the United States of America*, *97*(11), 6167–72.

Bruner, J. S. (1959). The Cognitive Consequences of Early Sensory Deprivation. *Psychosomatic Medicine*, *21*(2), 89–96.

Bu, C., Price, C. J., Frackowiak, R. S. J., & Friston, K. J. (1998). Different activation patterns in the visual cortex of late and congenitally blind subjects. *Brain*, *121*, 409–419.

Byrne, A., & Hilbert, D. R. (2003). Color realism and color science. *Behavioral and Brain Sciences*, *26*, 3–64.

Byrne, A., & Logue, H. (2009). *Disjunctivism: Contemporary Readings*. Cambridge (Mass): The MIT Press.

Chalmers, D. J. (1995). Facing Up to the Problem of Consciousness. *Journal of Consciousness Studies*, *2*(3), 200–219.

Chalmers, D. J. (2000). *What is a neural correlate of consciousness?* (T. Metzinger, Ed.) Cambridge (Mass): MIT Press.

Chalmers, D. J. (2006). Perception and the Fall from Eden. In T. Szabo Gendler & J. Hawthorne (Eds.), *Perceptual Experience* (pp. 49–125). Oxford: Oxford University Press.

Chapanis, N. P., Uematsu, S., Konigsmark, B. W., & Walker, A. (1973). Central phosphenes in man: a report of three cases. *Neuropsychologia*, *11*(1), 1–19.

Chekhov, A. (1904). *The Cherry Orchard*. http://www.gutenberg.org/ebooks/7986

Chemero, A. (2009). *Radical Embodied Cognitive Science*. Cambridge (Mass): MIT Press.

Chisholm, R. M. (1957). *Perceiving*. Ithaca: Cornell University Press.

Churchland, P. M. (2005). Chimerical colors: some phenomenological predictions from cognitive neuroscience. *Philosophical Psychology*, *18*(5), 527–560.

Clark, A. (2008). Supersizing the Mind. *8*(6), 79–99.

Clark, A., & Chalmers, D. J. (1998). The Extended Mind. *Analysis*, *58*(1), 10–23.

Cohen, J. (2009). *The Red and the Real: An Essay on Color Ontology*. New York: Oxford University Press.

Cohen, J., Hardin, C. L., & McLaughlin, B. P. (2006). True colours. *Analysis*, *66*(4), 335–40.

Cohen, L. G., Celnik, P., Pascual-Leone, A., Corwell, B., Falz, L., Dambrosia, J., … Hallett, M. (1997). Functional relevance of cross modal plasticity in blind humans. *Nature*, *389*(11), 180–182.

Collignon, O., & De Volder, A. G. (2009). Further evidence that congenitally blind participants react faster to auditory and tactile spatial targets. *Canadian Journal of Experimental Psychology, 63*(4), 287–93.

Collins, S. (2009). *Catching Fire [Hunger Games 2].* New York: Scholastic.

Coltheart, M. (1971). Visual feature-analyzers and after-effects of tilt and curvature. *Psychological Review, 78*(2), 114–21.

Conway, B. R. (2009). Color Vision, Cones, and Color-Coding in the Cortex. *The Neuroscientist, 15*(3), 274–290.

Conway, B. R., Kitaoka, A., Yazdanbakhsh, A., Pack, C. C., & Livingstone, M. S. (2005). Neural basis for a powerful static motion illusion. *The Journal of Neuroscience, 25*(23), 5651–6.

Cook, M. (1996). Descartes and the Dustbin of the Mind. *History of Philosophy Quarterly, 13*(1), 17–33.

Cooray, V., Cooray, G., & Dwyer, J. (2011). On the possibility of phosphenes being generated by the energetic radiation from lightning flashes and thunderstorms. *Physics Letters A, 375*(42), 3704–3709.

Cowey, A., & Walsh, V. Z. (2000). Magnetically induced phosphenes in sighted, blind and blindsighted observers. *Neuroreport, 11*(14), 3269–3273.

Craik, K. J. W. (1940). Origin of Visual After-Images. *Nature, 145,* 512.

Craik, K. J. W. (1966). On the effect of looking at the sun. In S. L. Sherwood (Ed.), *The Nature of Psychology* (pp. 98–101). Cambridge: Cambridge University Press.

Crane, H. D., & Piantanida, T. P. (1983). On seeing reddish green and yellowish blue. *Science, 221*(4615), 1078–80.

Crane, T. (2017), How we can be, *The Times Literary Supplement Limited,* https://www.the-tls.co.uk/articles/public/mind-body-problem-tim-crane/

Daniel, P. M., & Whitteridge, D. (1961). The Representation of the Visual Field on the Cerebral Cortex in Monkeys. *Journal of Physiology, 159,* 203–221.

Davidson, D. (1967). Causal Relations. *The Journal of Philosophy, 64*(21), 691–703.

Davidson, D. (1969). The Individuation of Events. In N. Rescher (Ed.), *In Essays in Honor of Carl G. Hempel: A Tribute in the Occasion of his Sixty-Fifth Birthday* (pp. 295–309). Dordrecht: D. Reidel Pub. Co.

Davidson, D. (1970). Events as Particulars. *Noûs, 4*(1), 25–32.

Davies, P. (2005). *About Time: Einstein's Unfinished Revolution.* New York: Simon & Schuster.

Davis, G., & Driver, J. (1994). Parallel detection of Kanizsa subjective figures in the human visual system. *Nature, 371,* 791–792.

Daw, N. W. (1962). Why After-Images Are Not Seen in Normal Circumstances. *Nature, 4860*(4), 1143–1145.

De Beni, R., & Cornoldi, C. (1988). Imagery limitations in totally congenitally blind subjects. *Journal of Experimental Psychology, 14*(4), 650–5.

Dennett, D. C. (1969). *Content and Consciousness.* London: Routledge & Kegan Paul.

Dennett, D. C. (1978). *Brainstorms: Philosophical Essays on Mind and Psychology.* (1st ed.). Montgomery: Bradford Books.

Dennett, D. C. (1991). *Consciousness Explained.* (1st ed.). Boston: Little Brown and Co.

Dennett, D. C., & Kinsbourne, M. (1992). Time and the Observer: the where and the when of Consciousness in the Brain. *Behavioral and Brain Sciences, 15*(1992), 183–247.

Dileep, G. (2008). *How the brain might work. A hierarchical and temporal model for learning and recognition.* Stanford University.

Dobelle, W. H., & Mladejovsky, M. G. (1974). Phosphenes Produced by Electrical Stimulation of Human Occipital Cortex, and their Application to the Development of a Prosthesis for the Blind. *Journal of Physiology, 243*, 553–576.

Dobelle, W. H., Mladejovsky, M. G., & Girvin, J. P. (1974). Artificial Vision for the Blind : Electrical Stimulation of Visual Cortex Offers Hope for a Functional Prosthesis. *Science, 183*(4123), 440–444.

Dolev, Y. (1989). A Real Present without Presentism 1. *Forthcoming, 4311*, 1–14.

Dolev, Y. (2007). *Time and Realism: Metaphysical and Antimetaphysical Perspectives. Book* (Vol. 69). Cambridge (Mass): MIT Press.

Domhoff, W. G. (2003). The Scientific Study of Dreams. Neural Networks, Cognitive Development, and Content Analysis. *II*(I), 13–33.

Dowe, P. (1992). Wesley Salmon's Process Theory of Causality and the Conserved Quantity Theory. *Philosophy of Science, 59*, 195–216.

Dowe, P. (2000). *Physical Causation.* New York: Cambridge University Press.

Dowe, P. (2007). *Causal Processes.* Stanford Encyclopedia of Philosophy.

Doya, K. (1999). What are the computations of the cerebellum, the basal ganglia and the cerebral cortex? *Neural Networks, 12*(7-8), 961–974.

Dretske, F. I. (1977). Causal Theories of Reference. *The Journal of Philosophy, 74*(10), 621. doi:10.2307/2025914

Dretske, F. I. (1995). *Naturalizing the Mind.* Cambridge (Mass): MIT Press.

Dretske, F. I. (2000). *Perception, Knowledge and Belief.* Cambridge: Cambridge University Press.

Driver, J., Haggard, P., & Shallice, T. (2007). *Mental processes in the human brain.* Oxford: Oxford University Press.

Dummett, M. (1978). *Truth and Other Enigmas.* London: Duckworth.

Ebersole, F. B. (1965). How Philosophers See Stars. *Mind, 74*(296), 509–529.

Eddington, A. S. (1929). *The Nature of the Physical World.* New York: MacMillan.

Einstein, A. (1916). *Relativity.* London: Routledge.

Eliot, T. S. (1935). *Burnt Norton.* New York: Harcourt Brace.

Farennikova, A. (2012). Seeing absence. *Philosophical Studies, 166*(3), 429–454.

Faubert, J., & Simon, H. A. (1999). The peripheral drift illusion: A motion illusion in the visual periphery. *Perception, 28*, 617–621.

Feigl, H. (1958). *The Mental and the Physical.* United States of America: University of Minnesota Press.

Feyerabend, P. K. (1975). *Against Method: Outline of an Anarchistic Theory of Knowledge.* London: NBL.

Ffytche, D. H. (2005). Visual hallucinations: Charles Bonnet syndrome. *Current Psychiatry Reports, 7*, 168–179.

Ffytche, D. H., Howard, R. J., Brammer, M. J., David, A., Woodruff, P., & Williams, S. (1998). The anatomy of conscious vision: an fMRI study of visual hallucinations. *Nature Neuroscience, 1*(8), 738–792.

Fischer, J., & Whitney, D. (2014). Serial dependence in visual perception. *Nature Neuroscience*, 1–9.

Fish, W. (2009). *Perception, Hallucination, and Illusion*. New York: Oxford University Press.

Fish, W. (2013). Perception, hallucination, and illusion: reply to my critics. *Philosophical Studies*, (163), 57–66.

Ford, L. S. (1974). The Duration of the Present. *Philosophy and Phenomenological Research*, 35(1), 100–106.

Foster, J. (1985). *Ayer*. London: Routledge.

Fraser, A. B., & Wilcox, K. J. (1979). Perception of illusory movement. *Nature, 281*, 565–566.

Furlong, E. J. (1951). *A Study in Memory*. London: Macmillan.

Gage, J. (1993). *Colour and Culture*. London: Thames & Hudson.

Gage, J. (2006). *Colour in Art*. London: Thames & Hudson.

Gale, R. M. (1971). Has the Present Any Duration? *Noûs, 5*(1), 39–47.

Galilei, G. (1623). *The Assayer*. https://web.stanford.edu/~jsabol/certainty/readings/Galileo-Assayer.pdf

Gallese, V. (2000). The Inner Sense of Action. *Journal of Consciousness Studies, 7*(10), 23–40.

Geisler, W. S. (1978). Adaptation, Afterimages and Cone Saturation. *Vision Research, 18*, 279–289.

Gerrits, H. J. M., DeHaan, B., & Vendrik, A. J. H. (1966). Experiments with retinal stabilized images: Relations between the observations and neural data. *Vision Research, 6*, 427–440.

Gerrits, H. J. M., & Vendrik, A. J. H. (1970). Simultaneous contrast, filling-in process and information processing in man's visual system. *Experimental Brain Research, 11*, 411–440.

Ghiradella, H. (1991). Light and color on the wing: butterflies and moths structural colors in. *Applied Optics, 30*(24), 3492–3501.

Gibson, J. J. (1979). *The Ecological Approach to Visual Perception*. Boston: Houghton Mifflin.

Gibson, J. J., & Radner, M. (1946). Adaptation, After-Effect and Constrast in the Perception of Tilted Lines. *Journal of Experimental Psychology, 20*, 453–467.

Gilroy, L. a, & Blake, R. (2005). The Interaction Between Binocular Rivalry and Negative Afterimages. *Current Biology, 15*(19), 1740–4. doi:10.1016/j.cub.2005.08.045

Goethe, J. W. (1810). *Theory of Colours*. London: Dover.

Gold, K., & Rabins, P. V. (1989). Isolated visual hallucinations and the Charles Bonnet syndrome: a review of the literature and presentation of six cases. *Comprehensive Psychiatry, 30*(1), 90–8.

Gothe, J., Brandt, S. a, Irlbacher, K., Röricht, S., Sabel, B. A., & Meyer, B.-U. (2002). Changes in visual cortex excitability in blind subjects as demonstrated by transcranial magnetic stimulation. *Brain, 125*(Pt 3), 479–90.

Grandin, T. (1996). *Thinking in Pictures: Other Reports from My Life with Autism*. New York: Basic Books.

Gregory, R. L., & Wallace, J. G. (1963). Recovery from early blindness: A case study. *Experimental Psychology Society Monograph, 2*(2), 1–44.

Grieve, K. L., Acuña, C., & Cudeiro, J. (2000). The primate pulvinar nuclei: vision and action. *Trends in Neurosciences, 23*(1), 35–9.

Haddock, A., & Macpherson, F. (2008). *Disjunctivism: Perception, Action, Knowledge.* (A. Haddock & F. Macpherson, Eds.) Oxford: Oxford University Press.

Haggard, P. (2011). Decision time for free will. *Neuron, 69*(3), 404–6.

Haggard, P., & Cole, J. (2007). Intention, attention and the temporal experience of action. *Consciousness and Cognition, 16*(2), 211–220.

Haggard, P., & Libet, B. (2001). Conscious Intention and Brain Activity. *Journal of Consciousness Studies, 8*(11), 47–63.

Haggard, P., Newman, C., & Magno, E. (1999). On the perceived time of voluntary actions. *The British Journal of Psychology, 90*(2), 191–303.

Hardin, C. L. (1993). *Color for Philosophers: Unweaving the Rainbow.* Indianapolis: Hackett.

Hardin, C. L. (2008). *Color Qualities and the Physical World.* (E. Wright, Ed.) Cambridge (Mass): MIT Press.

Harman, G. (1990). The Intrinsic Quality of Experience. *Philosophical Perspectives, 4*(1990), 31–52.

Harnad, S., & Scherzer, P. (2008). First, scale up to the robotic Turing test, then worry about feeling. *Artificial Intelligence in Medicine, 44*(2).

Hau, L. V. (1999). Light speed reduction to 17 metres per second in an ultracold atomic gas. *Nature, 397.6720,* 594–598.

Haynes, J.-D. (2009). Decoding visual consciousness from human brain signals. *Trends in Cognitive Sciences, 13*(5), 194–202.

Hedges, T. R. (2007). Charles Bonnet, his life, and his syndrome. *Survey of Ophthalmology, 52*(1), 111–4.

Herbert, F. (1965). *Dune.* New York: Chilton Books.

Hering, E. (1878). *Zur Lehre vom Lichtsinne.* Wien: C. Gerold's Son.

Hobson, A. J. (2002). *The Dream Drugstore. Chemically Altered States of Consciousness.* Cambridge (Mass): MIT Press.

Hobson, A. J. (2003). *Dreaming. A very short introduction.* Oxford: Oxford University Press.

Hofer, H., Singer, B., & Williams, D. R. (2005). Different sensations from cones with the same photopigment. *Journal of Vision, 5*(5), 444–454.

Hofmann, A. (1983). *LSD: My Problem Child.* Los Angeles: J. P. Tarcher.

Hogan, E. R., & English, E. a. (2012). Epilepsy and brain function: common ideas of Hughlings-Jackson and Wilder Penfield. *Epilepsy & Behavior, 24*(3), 311–3.

Holt, E. B. (1912). The Place of Illusory Experience in a Realistic World. In E. B. Holt, W. T. Marvin, W. P. Montague, R. B. Perry, W. B. Pitkin, & E. G. Spaulding (Eds.), *The New Realism* (pp. 303–373). New York: MacMillan Company.

Holt, E. B. (1914). *The Concept of Consciousness.* New York: MacMillan.

Holt, E. B., Marvin, W. T., Montague, W. P., Perry, R. B., Pitkin, W. B., & Spaulding, E. G. (1910). The Program and First Platform of Six Realists. *The Journal of Philosophy, 7*(15), 393–401.

Holt, E. B., Marvin, W. T., Montague, W. P., Perry, R. B., Pitkin, W. B., & Spaulding, E. G. (1912). *The New Realism.* New York: MacMillan Company.

Honderich, T. (1988). *The Consequences of Determinism. Book* (Vol. 2). Oxford: Clarendon Press.

Hopkins, R. (2000). Touching pictures. *The British Journal of Aesthetics, 40*(1), 149–167.

Horton, J. C., & Adams, D. L. (2005). The cortical column: a structure without a function. *Philosophical Transactions of the Royal Society of London B, 360*(1456), 837–862.

Hsieh, P. J., & Colas, J. T. (2012). Perceptual fading without retinal adaptation. *Journal of Experimental Psychology, 38*(2), 267–71.

Hsieh, P. J., & Tse, P. U. (2006). Illusory color mixing upon perceptual fading and filling-in does not result in "forbidden colors." *Vision Research, 46*(14), 2251–2258.

Huemer, M. (2001). *Skepticism and the Veil of Perception*. Cumnor Hill: Rowman and Littlefield.

Hume, D. (1758). *An Enquiry Concerning Human Understanding*. Chicago: Gateway.

Hurley, S. L. (1998). Vehicles, Contents, Conceptual Structure, and Externalism. *Analysis, 58*(1), 1–6.

Hurovitz, C. S., Dunn, S., Domhoff, W. G., & Fiss, H. (1999). The Dreams of Blind Men and Women: A Replication and Extension of Previous Findings. *Dreaming, 9*(2/3), 183–193.

Hurvich, L. M. (1981). *Color Vision*. Cambridge (Mass): Sinauer Associates Inc.

Hurvich, L. M., & Jameson, D. (1957). An Opponent-Process Theory of Color Vision. *Psychological Review, 64*(6), 384–404.

Huxley, A. (1954). *The Doors of Perception*. New York: Harper & Row.

Jackson, F., & Pargetter, R. (1977). Relative Simultaneity in the Special Relativity . *Philosophy of Science, 44*(3), 464–474.

Jackson, H. (1888). Epilepsy. *Brain, 11*, 179.

Jacob, P. (2008). *Ways of Seeing*. London: Oxford University Press.

James, W. (1890). *The Principles of Psychology*. New York: Henry Holt and Company.

James, W. (1904). Does "Consciousness" Exist? *The Journal of Philosophy, 1*(18), 477–491.

Johnston, M. (2001). Is Affect Always Mere Effect ? *Philosophy and Phenomenological Research, 63*(1), 225–228.

Johnston, M. (2004). The Obscure Object of Hallucination. *Philosophical Studies, 120*, 113–183.

Johnston, M. (2007). Objective Mind and the Objectivity of Our Minds. *Philosophy and Phenomenological Research, 75*(2), 233–268.

Jones, O. R. (1972). After-Images. *American Philosophical Quarterly, 9*(2), 150–158.

Kalderon, M. E. (2011). Color Illusion. *Noûs, 45*(4), 751–771.

Kammer, T. (1999). Phosphenes and transient scotomas induced by magnetic stimulation of the occipital lobe] their topographic relationship. *Neuropsychologia, 37*, 191–198.

Kammer, T., Puls, K., Strasburger, H., Hill, N. J., & Wichmann, F. a. (2005). Transcranial magnetic stimulation in the visual system. I. The psychophysics of visual suppression. *Experimental Brain Research, 160*(1), 118–128.

Kanai, R., Chaieb, L., Antal, A., Walsh, V., & Paulus, W. (2008). Frequency-Dependent Electrical Stimulation of the Visual Cortex. *Current Biology, 18*(23), 1839–1843.

Kanai, R., & Kamitani, Y. (2003). Time-locked perceptual fading induced by visual transients. *Journal of Cognitive Neuroscience, 15*(5), 664–72.

Kanai, R., Paulus, W., & Walsh, V. (2010). Transcranial alternating current stimulation (tACS) modulates cortical excitability as assessed by TMS-induced phosphene thresholds. *Clinical Neurophysiology, 121*(9), 1551–1554.

Kandinsky, W. (1911). *Concerning the Spiritual in Art*. The Floating Press.

Kanizsa, G. (1976). Subjective contours. *Scientific American, 234*, 48–52.

Kanizsa, G. (1991). *Vedere e pensare*. Bologna: Il Mulino.

Kanwisher, N. (2001). Neural events and perceptual awareness. *Cognition, 79*, 89–113.

Kar, K., & Krekelberg, B. (2012). Transcranial electrical stimulation over visual cortex evokes phosphenes with a retinal origin. *Journal of Neurophysiology, 108*(8), 2173–8.

Kaufman, D. M., & Solomon, S. G. (1992). Migraine visual auras. A medical update for the psychiatrist. *General Hospital Psychiatry, 14*(3), 162–70.

Kennedy, J. M. (1993). *Drawing and the Blind*. London: Yale University Press.

Kennedy, J. M., & Fox, N. (1977). Pictures to See and Pictures to Touch. In D. Perkins & B. Leondar (Eds.), *The Arts and Cognition* (pp. 118–135). London: John Hopkins University Press.

Kennedy, J. M., & Juricevic, I. (2003). Haptics and projection: Drawings by Tracy, a blind adult. *Perception, 32*, 1059–1071.

Kennedy, J. M., & Juricevic, I. (2006a). Blind man draws using diminution in three dimensions. *Psychonomic Bulletin and Review, 13*(3), 506–9.

Kennedy, J. M., & Juricevic, I. (2006b). *Esref Armagan and perspective in tactile pictures. Report.*

Kerr, N. H., & Domhoff, W. G. (2004). Do the Blind Literally "See" in Their Dreams? A Critique of a Recent Claim That They Do. *Dreaming, 14*(4), 230–233.

Kim, J. (1989). The Myth of Nonreductive Materialism. *Proceedings of the American Philosophical Society, 63*(3), 31–47.

Kim, J. (1993). The Non-Reductivist's Troubles with Mental Causation. In J. Heil & A. R. Mele (Eds.), *Mental Causation* (pp. 189–210). Oxford: Clarendon Press.

Kim, J. (1998). *Mind in a Physical World*. Cambridge (Mass): MIT Press.

Kim, J. (2005). *Physicalism, or Something Near Enough*. Princeton: Princeton University Press.

Kirshfeld, K. (1999). Afterimages: a tool for defining the neural correlate of visual consciousness. *Consciousness and Cognition, 8*(4), 462–83.

Kitaoka, A. (2003). Phenomenal characteristics of the peripheral drift illusion. *Vision, 15*, 261–262.

Klüver, H. (1966). *Mescal and Mechanisms of Hallucinations*. Chicago: Chicago University Press.

Koch, C. (2004). *The Quest for Consciousness: A Neurobiological Approach*. Englewood (Col): Roberts & Company Publishers.

Koch, C., & Crick, F. (2001). The zombie within. *Nature*, (411), 893.

Köhler, W., & Emery, D. A. (1947). Figural After-Effects in the Third Dimension of Visual Space. *The American Journal of Psychology, 60*(2), 159–201.

Kosslyn, S. M. (1996). *The Resolution of the Imagery Debate*. Cambridge (Mass): MIT Press.

Kosslyn, S. M., & Koenig, O. (1992). *Wet Mind. The New Cognitive Neuroscience*. New York: The Free Press.

Krill, A. E., Alpert, H. J., & Ostfeld, a M. (1963). Effects of a hallucinogenic agent in totally blind subjects. *Archives of Ophthalmology, 69*, 180–5.

Kunzendorf, R. G., Hartmann, E., Cohen, R., & Cutler, J. (1997). Bizarreness of the Dreams and Daydreams Reported by Individuals with Thin and Thick Boundaries. *Dreaming, 7*(4), 265–271.

Kupers, R., Fumal, A., Maertens de Noordhout, A., Gjedde, A., Schoenen, J., & Ptito, M. (2006). Transcranial magnetic stimulation of the visual cortex induces somatotopically organized qualia in blind subjects. *Proceedings of the National Academy of Sciences of the United States of America, 103*(35), 13256–13260.

Kuriki, I. (2008). Functional brain imaging of the Rotating Snakes illusion by fMRI. *Journal of Vision, 8*(10), 1–10.

Langsam, H. (2009). The Theory of Appearing Defended. In A. Byrne & H. Logue (Eds.), *Disjunctivism: Contemporary Readings* (pp. 181–207). Cambridge (Mass): MIT Press.

Le Poidevin, R. (2004). A Puzzle Concerning Time Perception. *Synthese, 142*(1), Le Poidevin, Robin. 2003. Travels in Four Dimensio.

Ledoux, J. E. (2012). Rethinking the emotional brain. *Neuron, 73*(4), 653–76.

Lennmarken, C., Bildfors, K., Enlund, G., Samuelsson, P., & Sandin, R. H. (2002). Victims of awareness. *Acta Anaesthesiologica Scandinavica, 46*(3), 229–31.

Lewis, D. K. (1986). *On the Plurality of Worlds*. London: Blackwell.

Lewis, D. K. (1991). *Parts of Classes*. Oxford: Blackwell.

Libet, B. (1981). The Experimental Evidence for Subjective Referral of a Sensory Experience Backwards in Time: Reply to P. S. Churchland. *Philosophy of Science, 48*(2), 182–197.

Libet, B. (1993). *Neurophysiology of Consciousness: Selected Papers and New Essays by Benjamin Libet*. Boston: Birkhauser.

Libet, B. (2002). The Timing of Mental Events: Libet's Experimental Findings and Their Implications. *Consciousness and Cognition, 11*(2), 291–299.

Libet, B. (2004). *Mind Time: The Temporal Factor in Consciousness*. Cambridge (Mass).

Libet, B., Wright, E. W., Feinstein, B., & Pearl, D. K. (1979). Subjective Referral of the Timing for a Conscious Sensory Experience. *Brain, 102*, 193–224.

Livitz, G., Yazdanbakhsh, A., Eskew, R. T., & Mingolla, E. (2011). Perceiving Opponents Hues in Color Induction Displays. *Seeing and Perceiving, 24*(1), 1–17.

Lockwood, P. L., Iannetti, G. D., & Haggard, P. (2013). Transcranial magnetic stimulation over human secondary somatosensory cortex disrupts perception of pain intensity. *Cortex, 49*(8), 2201–9.

Loewer, B. (2009). Why is there anything except physics? *Synthese, 170*(2), 217–233.

Logothetis, N. K. (2008). What we can do and what we cannot do with fMRI. *Nature, 453*(7197), 869–878.

Logothetis, N. K., & Wandell, B. a. (2004). Interpreting the BOLD signal. *Annual Review of Physiology, 66*, 735–69.

Lopes, D. M. (1997). Art Media and the Sense Modalities: Tactile Pictures. *The Philosophical Quarterly, 47*(189), 425–440.

Lopes, D. M. (2002). Vision, Touch, and the Value of Pictures. *The British Journal of Aesthetics, 42*(2), 191–201.

Lopes da Silva, F. H. (2003). Visual dreams in the congenitally blind? *Trends in Cognitive Sciences, 7*(8), 328–330.

Luna, L. E., & White, S. (2000). *Ayahuasca Reader: Encounters with the Amazon's Sacred Vine*. Santa Fe (New Mexico): Sinergetic Press.

Macpherson, F. (2013). The Philosophy and Psychology of Hallucination: An Introduction. In F. Macpherson & D. Platchias (Eds.), *Hallucination: Philosophy and Psychology* (pp. 1–38). Cambridge (Mass): MIT Press.

Magee, L. E., & Kennedy, J. M. (1980). Exploring pictures tactually. *Nature, 283*, 287–288.

Malebranche, N. (1923). *Dialogies on Metaphysics and Religion*. London: George Allen & Unwin.

Manzoni, A. (1840). *I Promessi Sposi*. Milano: Tipografia Guglielmini e Radaelli.

Manzotti, R. (2006a). A Process Oriented View of Conscious Perception. *Journal of Consciousness Studies, 13*(6), 7–41.

Manzotti, R. (2006b). Consciousness and existence as a process. *Mind and Matter, 4*(1), 7–43.

Manzotti, R. (2011a). The Spread Mind: Is Consciousness Situated? *Teorema, 30*(2), 55–78.

Manzotti, R. (2011b). The Spread Mind: Seven Steps to Situated Consciousness. *Journal of Cosmology, 14*, 4526–4541.

Manzotti, R. (2011c). The Spread Mind: Phenomenal Process-Oriented Vehicle Externalism. In M. Blamauer (Ed.), *The Mental as Fundamental. New Perspectives on Panpsychism.* Berlin: Ontos-Verlag.

Manzotti, R., & Moderato, P. (2013). Neuroscience: Dualism in Disguise. In A. Lavazza & H. Robinson (Eds.), *Contemporary Dualism* (pp. 81–97). New York: Routldege.

Manzotti, R., & Tagliasco, V. (2001). *Coscienza e Realtà. Una teoria della coscienza per costruttori e studiosi di menti e cervelli.* Bologna: Il Mulino.

Margalit, E., Maia, M., Weiland, J. D., Greenberg, R. J., Fujii, G. Y., Torres, G., ... Humayun, M. S. (2002). Retinal Prosthesis for the Blind. *Survey of Ophthalmology, 47*(4), 335–356.

Martin, M. G. F. (2002). The Transparency of Experience. *Mind and Language, 17*(4), 376–425.

Martin, M. G. F. (2004). The limits of self-awareness. *Philosophical Studies, 120*(37-89), 37–89.

Mather, G., Verstaten, F. A. J., & Anstis, S. (1998). The Motion After-Effect. *Trends in Cognitive Sciences, 2*(3), 111–117.

McGinn, C. (1999). *The Mysterious Flame: Conscious Minds in a Material World.* New York: Basic Books.

McTaggart, J. E. (1908). The Unreality of Time. *Mind, 17*(68), 457–474.

Mellor, D. H. (1998). *Real Time II.* London: Routldege.

Melzack, R., Israel, R., Lacroix, R., & Schultz, G. (1997). Phantom limbs in people with congenital limb deficiency or amputation in early childhood. *Brain, 120*, 1603–20.

Merabet, L. B., Battelli, L., Obretenova, S., Maguire, S., Meijer, P., & Pascual-Leone, A. (2009). Functional recruitment of visual cortex for sound encoded object identification in the blind. *Neuroreport, 20*(2), 132–8.

Merabet, L. B., Maguire, D., Warde, A., Alterescu, K., Stickgold, R. J., & Pascual-leone, A. (2004). Visual Hallucinations During Prolonged Blindfolding in Sighted Subjects. *Journal of Neuro-Ophthalmology, 24*(2), 109–113.

Merabet, L. B., & Pascual-Leone, A. (2010). Neural reorganization following sensory loss: the opportunity of change. *Nature Reviews Neuroscience, 11*(1), 44–53.

Merleau-Ponty, M. (1945). *The Phenomenology of Perception.* London: Routledge & Kegan Paul.

Merricks, T. (1999). Persistence, Parts, and Presentism. *Noûs, 33*(3), 421–438.

Merricks, T. (2001). *Objects and Persons.* Oxford: Oxford Clarendon Press.

Millikan, R. G. (1993). Content and vehicle. In N. Eilan, R. McCarthy, & B. Brewer (Eds.), *Spatial Representation.* Oxford: Blackwell.

Mollon, J. (1974). After-effects and the brain. *New Scientist, 61*(886), 479–482.

Molyneaux, B. (2009). Why Experience Told Me Nothing about Transparency. *Noûs, 43*(1), 116–136.

Montero, B. G. (2001). Post-Physicalism. *Journal of Consciousness Studies, 8*(2), 61–80.

Montero, B. G. (2013). Must Physicalism Imply the Supervenience of the Mental on the Physical? *The Journal of Philosophy, 5*, 93–110.

Moore, G. E. (1903). The Refutation of Idealism. *Mind*, *12*(48), 433–453.

Moore, G. E. (1912). *Lectures in Philosophy*. London: C. Lewy.

Mountcastle, V. B. (1997). The columnar organization of the neocortex. *Brain*, *120*(4), 701–722.

Murakami, H. (2002). *Kafka on the Shore*. New York: Vintage.

Murzyn, E. (2008). Do we only dream in colour? A comparison of reported dream colour in younger and older adults with different experiences of black and white media. *Consciousness and Cognition*, *17*, 1228–1237.

Myers, G. E. (1957). Perception and the "Time-Lag" Argument. *Analysis*, *17*(5), 97–102.

Newton, I, (1704). *Opticks: or, A treatise of the reflexions, refractions, inflexions and colours of light: Also two treatises of the species and magnitude of curvilinear figures*. London: Smith S. & Walford B.

Ney, A. (2010). Are There Fundamental Intrinsic Properties? In A. Hazlett (Ed.), *New Waves in Metaphysics (New Waves in Philosophy)*. Palgrave MacMillan.

Ney, A. (2015). Fundamental Physical Ontologies and the Constraint of Empirical Coherence: A Defense of Wave Function Realism. *Synthese*, *192* (10), 3105-3124.

Nida-Rumelin, M., & Suarez, J. (2009). Reddish Green: A Challenge for Modal Claims About Phenomenal Structure. *Philosophy and Phenomenological Research*, *78*(2), 346–391.

Noë, A. (2004). *Action in Perception*. Cambridge (Mass): The MIT Press.

Nudds, M. (2013). Naive Realism and Hallucination. In F. Macpherson & D. Platchias (Eds.), *Hallucination. Philosophy and Psychology* (pp. 271–290). Cambridge (Mass): MIT Press.

O'Regan, K. J. (1992). Solving the real misteries of visual perception: the world as an outside memory. *Canadian Journal of Psychology*, *46*(3), 461–488.

O'Regan, K. J. (2011). *Why Red Doesn't Sound Like a Bell. Understanding the Feel of Consciousness*. Oxford: Oxford University Press.

O'Regan, K. J., & Noë, A. (2001). A sensorimotor account of vision and visual consciousness. *Behavioral and Brain Sciences*, *24*(5), 939–73.

O'Regan, K. J., Rensink, R. A., & Clark, J. J. (1999). Change-blindness as a result of "mudsplashes". *Nature*, *398*(6722), 34.

Pascual-Leone, A., Amedi, A., Fregni, F., & Merabet, L. B. (2005). The Plastic Human Brain Cortex. *Annual Review of Neuroscience*, *28*, 377–402.

Paul, L. A., & Hall, N. (2013). *Causation: A User's Guide*. New York: Oxford University Press.

Paulus, W. (2010). On the difficulties of separating retinal from cortical origins of phosphenes when using transcranial alternating current stimulation (tACS). *Clinical Neurophysiology*, *121*(7), 987–91.

Pearl, J. (1998). *On the definition of actual cause*. Technical Report R-259, Department of Computer Science, University of California-Los Angeles, Los Angeles, CA.

Pearl, J. (2000). *Causality. Models, Reasoning, and Inference*. Cambridge: Cambridge University Press.

Penfield, W. (1950). *The Cerebral Cortex of Man*. New York.

Penfield, W. (1958). *The Excitable Cortex in Conscious Man*. Liverpool: Liverpool University Press.

Penfield, W. (1972). The electrode, the brain and the mind. *Zeitschrift Für Neurologie*, *201*(4), 297–309.

Penfield, W. (1975). *The Mystery of the Mind: A Critical Study of Consciousness and the Human Brain*. Princeton: Princeton University Press.

Penfield, W. (1996). Some Mechanisms of Consciousness Discovered During Electrical Stimulation of the Brain. *Proceedings of the National Academy of Sciences of the United States of America, 44*(6), 628–9.

Penfield, W., & Boldrey, E. (1937). Somatic Motor and Sensory Representation in the Cerebral Cortex of Man as Studied by Electrical Stimulation. *Brain,* 389–434.

Penfield, W., & Perot, P. (1963). The Brain's Record of Auditory and Visual Experience: A Final Summary and Discussion. *Brain, 86*(4), 595–696.

Penfield, W., & Rasmussen, T. (1950). *The Cerebral Cortex of Man. A Clinical Study of Localization of Function.* New York: MacMillan Company.

Penrose, R. (1989). *The Emperor's New Mind.* Oxford: Oxford University Press.

Pessoa, L., Thompson, E., & Noë, A. (1998). Finding out about filling-in: A guide to perceptual completion for visual science and the philosophy of perception. *Behavioral and Brain Sciences, 21,* 723–802.

Phillips, I. (2013). Afterimages and Sensation. *Philosophy and Phenomenological Research, 87*(2), 417–453.

Pietrini, P., Furey, M. L., Ricciardi, E., Gobbini, I. M., Wu, W.-H. C., Cohen, L. G., … Haxby, J. V. (2004). Beyond sensory images: Object-based representation in the human ventral pathway. *Proceedings of the National Academy of Sciences of the United States of America, 101*(15), 5658–63.

Pietrobon, D., & Striessnig, J. (2003). Neurobiology of migraine. *Nature Reviews Neuroscience, 4*(5), 386–98.

Place, U. T. (1956). Is consciousness a brain process? *The British Journal of Psychology, 47,* 44–50.

Poggio, T., & Torre, V. (1990). *Ill-Posed Problems and Regularization Analysis in Early Vision.* MIT AI Lab.

Pollen, D. A. (2004). Brain stimulation and conscious experience. *Consciousness and Cognition, 13*(3), 626–45.

Pollen, D. A. (2006). Brain stimulation and conscious experience: electrical stimulation of the cortical surface at a threshold current evokes sustained neuronal activity only after a prolonged latency. *Consciousness and Cognition, 15*(3), 560–565.

Pons, T. (1996). Novel sensations in the congenitally blind. *Nature, 380,* 479–480.

Power, S. E. (2010). Perceiving External Things and the Time-Lag Argument. *European Journal of Philosophy, 21*(1), 94–117.

Power, S. E. (2011). The Metaphysics of the "Specious" Present. *Erkenntnis, 77*(1), 121–132.

Ptito, M., Fumal, A., de Noordhout, a M., Schoenen, J., Gjedde, A., & Kupers, R. (2008). TMS of the occipital cortex induces tactile sensations in the fingers of blind Braille readers. *Experimental Brain Research, 184*(2), 193–200.

Ptito, M., Kupers, R., Lomber, S., & Pietrini, P. (2012). Sensory deprivation and brain plasticity. *Neural Plasticity, 2012,* 810370.

Purves, D., & Beau Lotto, R. (2002). The empirical basis of color perception. *Consciousness and Cognition, 11*(4), 609–629.

Putnam, H. (1967). Time and Physical Geometry. *The Journal of Philosophy, 64*(8), 240–247.

Ramachandran, V. S., & Gregory, R. L. (1991). Perceptual filling in of artificially induced scotomas in human vision. *Nature, 350*(6320), 699–702.

Ramachandran, V. S., & Hubbard, E. M. (2001). Synaesthesia: A windows into perception, thought and language. *Journal of Consciousness Studies, 8*(12), 3–34.

Ramachandran, V. S., & McGeoch, P. D. (2007). Occurrence of phantom genitalia after gender reassignment surgery. *Medical Hypotheses, 69*(5), 1001–3.

Ramachandran, V. S., & McGeoch, P. D. (2008). Phantom Penises In Transsexuals: Evidence of an Innate Gender-Specific Body Image in the Brain. *Journal of Consciousness Studies, 15*(1), 3–26.

Reber, A. S. (1992). The Cognitive Unconscious: An Evolutionary Perspective. *Consciousness and Cognition, 1,* 93–133.

Reddy, L., & Kanwisher, N. (2006). Coding of visual objects in the ventral stream. *Current Opinion in Neurobiology, 16*(4), 408–414.

Reichenbach, H. (1958). *The Philosophy of Space and Time.* New York: Dover.

Reichenbach, H. (1971). *The Direction of Time.* Berkeley: University of California Press.

Revonsuo, A., & Salmivalli, C. (1995). A content analysis of bizarre elements in dreams. *Dreaming, 5*(3), 169–187.

Rizzolatti, G., Fogassi, L., & Gallese, V. (2001). Neurophysiological mechanisms underlying the understanding and imitation of action. *Nature Reviews Neuroscience, 2,* 661–670.

Robinson, H. (1994). *Perception.* London: Routledge.

Rockwell, T. (2005). *Neither Ghost Nor Brain.* Cambridge (Mass): MIT Press.

Russell, B. (1912a). On The Notion of Cause. *Proceedings of the Aristotelian Society, 13,* 1–26.

Russell, B. (1912b). *The Problems of Philosophy.* London: T. Butterworth.

Russell, B. (1927). *The Analysis of Matter.* London: Routledge & Kegan Paul.

Russell, B. (1948). *Human Knowledge, Its Scope and Limits.* New York: Simon & Schuster.

Ruzzoli, M., Gori, S., Pavan, A., Pirulli, C., Marzi, C. a, & Miniussi, C. (2011). The neural basis of the Enigma illusion: a transcranial magnetic stimulation study. *Neuropsychologia, 49*(13), 3648–55.

Saadah, E. S., & Melzack, R. (1994). Phantom limb experiences in congenital limb-deficient adults. *Cortex, 30*(3), 479–485.

Sacks, O. (1970). *Migraine.* Berkeley: University of California Press.

Sacks, O. (2012). *Hallucinations.* Canada: Alfred E. Knopf.

Sadato, N., Pascual-Leone, A., Grafman, J., Ibanez, V., Deiber, M.-P., Dold, G., & Hallett, M. (1996). Activation of the primary visual cortex by Braille reading in blind subjects. *Nature, 380,* 526–528.

Salminen-Vaparanta, N., Vanni, S., Noreika, V., Valiulis, V., Móró, L., & Revonsuo, A. (2013). Subjective Characteristics of TMS-Induced Phosphenes Originating in Human V1 and V2. *Cerebral Cortex,* 1–10.

Salmon, W. C. (1969). The Conventionality of Simultaneity. *Philosophy of Science, 36*(1), 44–63.

Salmon, W. C. (1997). Causality and Explanation. *Philosophy of Science, 64*(3), 461–477.

Sayin, H. Ü. (2014). Does the Nervous System Have an Intrinsic Archaic Language ? Entoptic Images and Phosphenes. *Neuroquantology, 12*(3), 427–445.

Schaal, D. W. (2005). Naming Our Concerns about Neuroscience: A Review of Bennett and Hacker's Philosophical Foundations of Neuroscience. *Journal of the Experimental Analysis of Behavior, 84*(3), 683–692.

Schaffer, J. (2014). The Metaphysics of Causation. *Stanford Encyclopedia of Philosophy.*

Schellenberg, S. (2010). The particularity and phenomenology of perceptual experience. *Philosophical Studies, 149*(1), 19–48.

Schellenberg, S. (2011). Ontological Minimalism about Phenomenology. *Philosophy and Phenomenological Research, 83*(1), 1–40.

Schlesinger, G. N. (1970). Change and Time. *The Journal of Philosophy, 67*(9), 294–300.

Schutter, D. J. L. G., & Hortensius, R. (2010). Retinal origin of phosphenes to transcranial alternating current stimulation. *Clinical Neurophysiology, 121*(7), 1080–4.

Schwitzgebel, E. (2002a). How Well Do We Know Our Own Conscious Experience? The Case of Visual Imagery. *Journal of Consciousness Studies, 9*(5-6), 35–53.

Schwitzgebel, E. (2002b). Why did we think we dreamed in black and white? *Studies in History and Philosophy of Science, 33*(4), 649–660.

Schwitzgebel, E. (2007). Do You Have Constant Tactile Experience of Your Feet in Your Shoes? Or Is Experience Limited to What's in Attention? *Journal of Consciousness Studies, 14*(3), 5–35.

Schwitzgebel, E. (2008). The Unreliability of Naive Introspection. *Philosophical Review, 117*(2), 245–273.

Schwitzgebel, E., Huang, C., & Zhou, Y. (2006). Do we dream in color? Cultural variations and skepticism. *Dreaming, 16*(1), 36–42.

Searle, J. R. (1984). *Minds, Brains, and Science.* Cambridge (Mass): Harvard University Press.

Searle, J. R. (1992). *The Rediscovery of the Mind.* Cambridge (Mass): MIT Press.

Sekuler, R. W., & Ganz, L. (1963). After-effect of seen motion with a stabilized image. *Science, 139,* 419–420.

Sellars, W. (1960). Philosophy and the Scientific Image of Man. In R. Colodny (Ed.), *Frontiers of Science and Philosophy* (pp. 1–40). Pittsburgh: University of Pittsburgh Press.

Senden, V. (1932). *Raum und Gestaltauffassung bei operierten Blindgeborenen vor und nach der Operation (conception of space and gestalt in congenital blind children before and after surgery).* Leipzig: Verlag Jd Ambrosus Barth.

Seymour, B., & Dolan, R. (2008). Emotion, decision making, and the amygdala. *Neuron, 58*(5), 662–671.

Shams, P. N., & Plant, G. T. (2011). Migraine-like visual aura due to focal cerebral lesions: case series and review. *Survey of Ophthalmology, 56*(2), 135–61.

Shanon, B. (2002). Ayahuasca Visualizations: A Structural Typology. *Journal of Consciousness Studies, 9*(2), 3–30.

Shanon, B. (2010). The epistemics of ayahuasca visions. *Phenomenology and the Cognitive Sciences, 9,* 263–280.

Shapley, R. (2002). Neural mechanisms for color perception in the primary visual cortex. *Current Opinion in Neurobiology, 12*(4), 426–432.

Shapley, R., & Hawken, M. J. (2011). Color in the cortex: single- and double-opponent cells. *Vision Research, 51*(7), 701–17.

Shepherd, R. K., Shivdasani, M. N., Nayagam, D. a X., Williams, C. E., & Blamey, P. J. (2013). Visual prostheses for the blind. *Trends in Biotechnology, 31*(10), 562–71.

Shimojo, S., Kamitani, Y., & Nishida, S. (2001). Afterimage of Perceptually Filled-In Surface. *Science, 293*(August), 1677–1681.

Shoemaker, S. S. (1969). Time Without Change. *The Journal of Philosophy, 66*(12), 363–381.

Shoemaker, S. S. (1990). Qualities and Qualia: What's in the Mind? *Philosophy and Phenomenological Research, 50*(May), 109–131.

Shum, D. (1999). Implicit and Explicit Memory in Children with Traumatic Brain Injury. *Journal of Clinical and Experimental Neuropsychology (Neuropsychology, Development and Cognition: Section A), 21*(2), 149–158.

Sider, T. (2001). *Four-Dimensionalism.* Oxford: Oxford Clarendon Press.

Sider, T., & Braun, D. (2007). Vague, so Untrue. *Noûs, 41,* 133–146.

Siegel, K. (1978). Cocaine Hallucinations. *American Journal of Psychiatry, 135,* 309–314.

Siegel, S. (2010). *The Contents of Visual Experience.* Oxford: Oxford University Press.

Simons, D. J. (2000). Attentional capture and inattentional blindness. *Trends in Cognitive Sciences, 4,* 147–155.

Simons, D. J., & Chabris, C. F. (1999). Gorillas in our midst: Sustained inattentional blindness for dynamic events. *Perception, 28*(28), 1059–1074.

Simons, D. J., Lleras, A., Martinez-Conde, S., Slichter, D., Caddigan, E., & Nevarez, G. (2006). Induced visual fading of complex images. *Journal of Vision, 6*(10), 1093–101.

Simons, P. M. (1987). *Parts: A Study in Ontology.* Oxford: Clarendon Press.

Skow, B. (2010). On the meaning of the question "How fast does time pass?" *Philosophical Studies, 155*(3), 325–344.

Skow, B. (2011). Experience and the Passage of Time. *Philosophical Perspectives, 25,* 359–387.

Smart, J. J. C. (1959). Sensations and Brain Processes. *The Philosophical Review, 68*(2), 141–156.

Smart, J. J. C. (1963). *Philosophy and Scientific Realism.* London: Routledge & Kegan Paul.

Smith, D. A. (2002). *The Problem of Perception.* Cambridge (Mass): Harvard University Press.

Smolin, L. (2013). Temporal naturalism. *arXiv Preprint,* 1–42.

Spillmann, L., Otte, T., Hamburger, K., & Magnussen, S. (2006). Perceptual filling-in from the edge of the blind spot. *Vision Research, 46*(25), 4252–4257.

Stevens, W. (1971). *The Collected Poems.* New York: Alfred E. Knopf.

Stoerig, P. (1996). Varieties of vision: from blind responses to conscious recognition. *Trends In Neurosciences, 19*(9), 401–406.

Stokes, J. (2005). The blind painter and the Cartesian Theater.

Strawson, G. (2005). *Why physicalism entails panpsychism.* Danish National Research Foundation.

Strawson, G. (2008). *Real Materialism and Other Essays.* Oxford: Oxford Clarendon Press.

Suchting, W. A. (1969). Perception and the Time-Gap Argument. *The Philosophical Quarterly, 19*(74), 46–56.

Tanaka, K. (1997). Columnar organization in the inferotemporal cortex. *Cerebral Cortex, 12,* 469–498.

Taylor, J. G. (2001). *The Race for Consciousness.* London: Bradford Books.

Thompson, E. (2007). *Mind in Life: Biology, Phenomenology, and the Sciences of Mind.* Cambridge (Mass): The Belknap Press of the Harvard University Press.

Thompson, E., & Varela, F. J. (2001). Radical embodiment: neural dynamics and consciousness. *Trends in Cognitive Sciences, 5*(10), 418–425.

Tong, F. (2003). Primary visual cortex and visual awareness. *Nature Reviews Neuroscience*, *4*(3), 219–229.

Tong, F., & Pratte, M. S. (2012). Decoding patterns of human brain activity. *Annual Review of Psychology*, *63*, 483–509.

Tonneau, F. (2004). Consciousness Outside the Head. *Behavior and Philosophy*, *32*(February 2002), 97–123.

Tooley, M. (1997). *Time, Tense, and Causation*. Oxford: Oxford University Press.

Tootell, R. B. H., Silverman, M. S., Switkes, E., & De Valois, R. L. (1982). Deoxyglucose Analysis or Retinotopic Organization in Primate Striate Cortex. *Science*, *218*(4575), 901–903.

Travis, C. (2004). The Silence of the Senses. *Mind*, *113*(1), 449–467.

Trimble, M. R. (2007). *The Soul in the Brain: The Cerebral Basis of Language, Art, and Belief*. Baltimore: John Hopkins University Press.

Tsuchiya, N., & Koch, C. (2005). Continuous flash suppression reduces negative afterimages. *Nature Neuroscience*, *8*(8), 1096–1101.

Tye, M. (2002). Representationalism an the Transparency of Experience. *Noûs*, *36*(1), 137–151.

Tye, M. (2010). The Puzzle of Transparency. In A. Byrne, J. Cohen, G. Rosen, & S. Shiffrin (Eds.), *The Norton Introduction to Philosophy*. London: Norton.

Uttal, W. R. (2001). *The New Phrenology: The Limits of Localizing Cognitive Processes in the Brain*. Boston: MIT Press.

Valberg, J. J. (1992). *The Puzzle of Experience*. Oxford: Clarendon Press.

Van Boxtel, J. J. A., Tsuchiya, N., & Koch, C. (2010). Opposing effects of attention and consciousness on afterimages. *Proceedings of the National Academy of Sciences of the United States of America*, *107*(19), 8883–8.

Van Inwagen, P. (1990a). Four-Dimensional Objects. *Noûs*, *24*(2), 245–255.

Van Inwagen, P. (1990b). *Material Beings. Book* (Vol. 53). New York: Cornell University Press.

Van Inwagen, P., & Zimmerman, D. W. (2007). *Persons. Human and Divine.* (P. van Inwagen & D. W. Zimmerman, Eds.) Oxford: Oxford University Press.

Vandenbos, G. R. (2007). *APA Dictionary of Psychology.* (G. R. Vandenbos, Ed.)Washington (DC): American Psychological Association.

Varela, F. J. (1999). *The Specious Present: A Neurophenomenology of Time Consciousness.* (J. Petitot, F. Varela, P. Pachould, & J. M. Roy, Eds.) Stanford (Cal): Stanford University Press.

Virsu, V., & Laurinen, P. (1977a). Long-Lasting Afterimages Caused by Neural Adapatation. *Vision Research*, 853–860.

Virsu, V., & Laurinen, P. (1977b). Long-Lasting Afterimages Caused by Neural Adaptation. *Vision Research*, *17*, 853–860.

Vlasov, Y. A., O'Boyle, M., Hamann, H. F., & McNab, S. J. (2005). Active control of slow light on a chip with photonic crystal waveguides. *Nature*, *438*(7064), 65–9.

Von Campenhausen, C., & Schramme, J. (1995). 100 Years of Benham's Top in Colour Science. *Perception*, *24*(6), 695–717.

Von Uexküll, J. (1909). *Umwelt und Innenwelt der Tiere*. Berlin: Springer.

Von Uexküll, J. (1957). A Stroll Through the Worlds of Animals and Men. In C. H. Schiller (Ed.), *Instinctive Behavior. The Development of a Modern Concept* (pp. 5–80). New York: International University Press.

Vukusic, P., & Sambles, J. R. (2003). Photonic structures in biology. *Nature, 424*(6950), 852–5.

Vukusic, P., Sambles, J. R., Lawrence, C. R., & Wootton, R. J. (2001). Now you see it—now you don't. *Nature, 410*(303), 36.

Wachowski, A., & Wachowski, L. (1999). *The Matrix. Movie.*

Weiskrantz, L. (1990). *Blindsight: A Case Study and Implications.* Oxford: Clarendon Press.

Wheeler, R. H. (1918). Visual Phenomena in the Dreams of a Blind Subject. *Psychological Review, 18*(3), 315–312.

Whitehead, A. N. (1925a). *Science and Philosophy.* New York: Philosophical Library.

Whitehead, A. N. (1925b). *Science and the Modern World.* New York: Macmillan Company.

Whitehead, A. N. (1929). *Process and Reality.* London: Free Press.

Whitehead, A. N. (1933). *Adventures of Ideas.* New York: Free Press.

Winawer, J., Juk, A. C., & Boroditsky, L. (2008). A Motion Aftereffect From Still Photographs Depicting Motion. *Psychological Science, 19*(3), 276–282.

Woolf, V. (1923). *Mrs. Dalloway. Book.*

Yablo, S. (2004). Advertisement for a Sketch of an Outline of a Prototheory of Causation. In N. Hall, L. A. Paul, & J. Collins (Eds.), *Causation and Counterfactuals* (pp. 119–137). Cambridge (Mass): MIT Press.

Yarrow, K., Haggard, P., & Rothwell, J. C. (2004). Action, arousal, and subjective time. *Consciousness and Cognition, 13*(2), 373–390.

Zaidi, Q., Ennis, R., Cao, D., & Lee, B. B. (2012). Neural locus of color afterimages. *Current Biology, 22*(3), 220–4.

Zajonc, A. (1995). *Catching the Light.* Oxford: Oxford University Press.

Zeki, S. (1973). Colour coding in rhesus monkey prestriate cortex. *Brain Research, 53*(2), 422–7.

Zeki, S. (1978). Functional specialization in the visual cortex of the rhesus monkey. *Nature,* (274), 423–428.

Zeki, S. (2001). Localization and Globalization in Conscious Vision. *Annual Review of Neuroscience, 24*, 57–86.

Zeki, S., & Moutoussis, K. (1997). Temporal hierarchy of the visual perceptive systems in the Mondrian world. *Philosophical Transactions of the Royal Society of London B, 264*, 1415–1419.

Zeki, S., Watson, J. D. G., & Frackowiak, R. S. J. (1993). Going beyond the information given: the relation of illusory visual motion to brain activity. *Philosophical Transactions of the Royal Society of London B, 252*(1335), 215–22.

Zubek, J. P. (1969). *Sensory Deprivation: Fifteen Years of Research.* New York: Appleton-Century-Crofts.

ENDNOTES

1   Beginning in 2011, I used the expression "the spread mind" thanks to a friendly suggestion from the novelist Tim Parks. The term is unrelated to Clarks and Chalmers' "extended mind." Clark & Chalmers, 1998; Clark, 2008. To my chagrin, some earlier works of mine are not representative of the view put forward in this book. Manzotti, 2011a, 2011b, 2011c.

2   Outstanding experimental results have shown that it is possible to reconstruct one's experience and mental content from neural recording. For instance, Haynes, 2009; Tong & Pratte, 2012; Tong, 2003. Haynes and Rees, for example, in 2006 succeeded in matching specific brain activity with specific visual experience. More remarkably, in 2011 Nishimoto managed to reconstruct the external visual stimuli that volunteers were responding to on the basis of their brain activity. Yet, such techniques do not really *read minds* or *visualize one's thoughts and experiences.* They are statistical predictions about one's thoughts based on recurrent neural patterns. In fact, they correlate as much with the external causes.

3   Brewer, 2011; Harman, 1990.

4   In this book, *internal* and *external* have no metaphysical implications. They mean inside or outside one's body and, when specified, inside or outside one's brain. Neural firings are internal, while a red apple on the table is external. Hybrid cases such as visceral pain are skipped.

5   Lately, many authors have pointed out the difficulty of providing a positive account of the physical. Kim, 1993, 2005; Loewer, 2009; Montero, 2001, 2013; Strawson, 2005.

6   Chemero, 2009; Noë, 2004; O'Regan & Noë, 2001; O'Regan, 2011; Rockwell, 2005; Thompson & Varela, 2001; Thompson, 2007.

7   The idea of spread mind is not a panpsychist view. It does not endorse a mental world smeared over the physical one. Rather, spread mind sweeps away the need to add an additional phenomenal level to the physical world. Neither is the view akin to Russell's neutral monism or to James's doctrine of pure experience. Spread mind does not flesh the world out of a metaphysically neutral stuff. Physical objects and causal processes are all that is required.

8   Famously, bishop George Berkeley 1685-1753 claimed that objects exist only insofar as someone perceives them. He contended that familiar objects like apples are only ideas in the minds of perceivers.

9   When I use the expression "to bring something into existence," I mean enabling something to take place.

10   Cook, 1996; Shoemaker, 1990.

11   Of late, the notion of look has been questioned by Charles Travis, 2004.

12   Cook, 1996; Shoemaker, 1990.

13   Shangri-La is different from Chalmers's Eden, where metaphysical concessions are neces-
     sary to allow direct grasping of intrinsic properties of things. Shangri-La is a nomological
     possibility. It is a city that could exist in our world and that would not require any modifi-
     cation of known laws of nature.

14   The heated debate as to the nature of the physical is not crucial here. Montero, 2001; Ney,
     2010, 2015; Strawson, 2008.

15   Of course, I have no pretense of historical rigor here. The Galilean object is a rhetorical
     device I deploy to pin down the notion of an object different from what one experiences.

16   Here "absolute" is used in its etymological sense of being *ab+solute*, namely something
     that is what it is independently from anything else, something whose nature is cut off
     from the rest.

17   Of course, much better physicalist models of colors are available. Byrne & Hilbert, 2003;
     J. Cohen, Hardin, & McLaughlin, 2006; J. Cohen, 2009; Hardin, 1993. For the sake of the
     discussion, I use frequency.

18   In this regard, Wilfred Sellars contrasted the manifest and scientific image of the world.
     Sellars, 1960.

19   Science and philosophy—although each jealous of the other's methods and topics—are
     not necessarily different fields of enquiry. Whitehead, 1925a, 1933.

20   A perfect example of this approach is offered by Eric Schwitzgebel. It is enough to read the
     titles of his works to get a clear idea: Schwitzgebel E. 2002a. *How Well Do We Know Our
     Own Conscious Experience?* The Case of Visual Imagery. Journal of Consciousness Studies,
     95-6, 35–53.; Schwitzgebel, E. 2002b. *Why did we think we dreamed in black and white?*
     Studies in History and Philosophy of Science, 334, 649–660; Schwitzgebel, E. 2008. *The
     Unreliability of Naive Introspection.* Philosophical Review, 1172, 245–273.

21   The causal root of the object is such that it makes sense to ask a further question about
     the spread object—namely, when does it take place? In fact, for causal process I refer to
     an actual physical process in the sense outlined—among others—by Arthur Reichenbach
     or Phil Dowe. Dowe, 2000; Reichenbach, 1958. A causal process is a process transferring
     some physical qantity. Pseudo-processes are not considered.

22   God's view is a metaphor that refers to an all-encompassing angle on physical reality.

23   It is revealing that in English, *immaterial* means lack of causal relevance.

24   I already presented this example in an Italian work I wrote many years ago: Manzotti &
     Tagliasco, 2001.

25   The reply to Molineux's question is of course negative since, apart from contingent simi-
     larities, nothing conjoins the tactile physical phenomena and the visual physical phenom-
     ena or any other combination to be alike. Of course, sometimes, commonalities might
     occur. Such cases do not need to be the expression of an innate or *a priori* phenomenology.

26 Consider light. Light rays can be split based on their spatial frequency. Thus, the high frequency blurring resulting from common myopia results from filtering out the highest frequencies of light. In other words, when I take off my glasses, I set aside a portion of the environment not unlike what I do if I occlude a part of an object.

27 Bartels & Zeki, 1998, 2005; Zeki & Moutoussis, 1997; Zeki, 1978.

28 Until 2014, I was a fervent supporter of process ontology: Manzotti, 2006a, 2006b, 2011b. It was a mistake.

29 In this regard, James claimed that according to Berkeley, "what common sense means by realities is exactly what the philosopher means by ideas" (James, 1904, p. 481). It is relevant here, since James outlined his version of neutral monism. Here I do not attempt to do justice to Berkeley's view.

30 From the viewpoint of an evolutionist, this is a suspicious explanation. If perception were reliable hallucination, one should concede that, first, animals developed the capability of hallucinating, and then learned to tune hallucinations to match them with the actual world. Hardly plausible.

31 https://www.ted.com/talks/david_chalmers_how_do_you_explain_consciousness/transcript?language=en

32 The notion of causal geometry captures nicely the gist of what follows, since it suggests a sort of causal hylomorphism (every physical manifestation is the sum of what has gone before). It is a nice shorthand to refer to how causation carves out the world we perceive.

33 This is the time-gap argument turned upside down: Suchting, 1969. More on this will follow.

34 Bartels & Zeki, 2005; Libet, 2004; Reddy & Kanwisher, 2006.

35 Bartels & Zeki, 2004, 2005.

36 Ebersole, 1965.

37 Ayer, 1956, pp. 93–95, 156.

38 Dolev, 2007.

39 Aleman, van Lee, Mantione, Verkoijen, & de Haan, 2001; Arditi, Holtzman, & Kosslyn, 1988; Bu, Price, Frackowiak, & Friston, 1998; Collignon & De Volder, 2009; De Beni & Cornoldi, 1988; Lopes da Silva, 2003; Pons, 1996.

40 Brugger et al., 2000; Churchland, 2005; Kennedy, 1993; Ramachandran & McGeoch, 2008; Saadah & Melzack, 1994.

41 Kunzendorf, Hartmann, Cohen, & Cutler, 1997; Murzyn, 2008; Revonsuo & Salmivalli, 1995; Schwitzgebel, 2002a.

42 Brindley & Lewin, 1968; Dobelle & Mladejovsky, 1974; Kammer, 1999.

43 Block, 2005b.

44 Kupers et al., 2006.

45 Schwitzgebel, Huang, & Zhou, 2006; Schwitzgebel, 2002b.

46 Schwitzgebel, 2002a, 2008.

47 Albus, 2000; Byrne & Hilbert, 2003; Gage, 1993, 2006; Hardin, 1993, 2008.

48 O'Regan & Noë, 2001; O'Regan, 1992; Poggio & Torre, 1990.

49 Faubert & Simon, 1999; Fraser & Wilcox, 1979; Kitaoka, 2003.

50 Conway, Kitaoka, Yazdanbakhsh, Pack, & Livingstone, 2005; Kitaoka, 2003; Kuriki, 2008.

51 Alleysson & Méary, 2012; Davis & Driver, 1994; Kanizsa, 1976, 1991.

52 Albus, 2000; Gage, 1993; Goethe, 1810; Hardin, 1993; Kandinsky, 1911; Newton, 1704.

53 Isaac Newton himself criticized such an over-simplistic color model. Notwithstanding many excellent analyses of color in terms of physical properties (Byrne & Hilbert, 2003), the identity between colors and light frequency is still surprisingly popular.

54 Ghiradella, 1991; Vukusic, Sambles, Lawrence, & Wootton, 2001; Vukusic & Sambles, 2003.

55 Aleman & Larøi, 2008; Allen, Larøi, McGuire, & Aleman, 2008; Arieti, 1974; Association, 1975; Berrios & Markova, 2015; Blom, 2010; Broad, 1952; Foster, 1985; Russell, 1912b; S. Siegel, 2010; Vandenbos, 2007.

56 Penfield, 1950, 1958, 1975.

57 Chalmers, 2000; Dennett, 1969; Jacob, 2008; Kosslyn & Koenig, 1992; Kosslyn, 1996; Penrose, 1989; Searle, 1992; Strawson, 2008; Taylor, 2001; Tong, 2003; Trimble, 2007; van Inwagen & Zimmerman, 2007.

58 Traditional cases of stimulation have been reported by Borchers, Himmelbach, Logothetis, & Karnath, 2012; Brindley & Lewin, 1968; Penfield & Rasmussen, 1950; Pollen, 2004. More recent cases of less-invasive brain stimulation have been studied by Lockwood, Iannetti, & Haggard, 2013; Pollen, 2004, 2006; Ptito et al., 2008; Salminen-Vaparanta et al., 2013.

59 Hobson, 2002; Hoffman, 1983; Huxley, 1954; Luna & White, 2000; Shanon, 2002.

60 Hobson, 2002, 2003; K. Siegel, 1978.

61 Dowe, 2007; Reichenbach, 1958; Salmon, 1997.

62 Merabet et al., 2004.

63 Ffytche, 2005; Ffytche et al., 1998; Gold & Rabins, 1989; Hedges, 2007.

64 Bruner, 1959; Ptito, Kupers, Lomber, & Pietrini, 2012; Zubek, 1969.

65 Hobson, 2002; Shanon, 2010.

66 Not completely though. There are many possible spatial and temporal ways to reshuffle the world due to the ways in which bodies can allow objects to take place: e.g., the Thatcher effect, temporal displacement, or subjective referral in time. Haggard, Newman, & Magno, 1999; Libet, Wright, Feinstein, & Pearl, 1979.

67 As regards congenitally blind subjects, the only documented case is that of the Turkish painter as described by Kennedy & Juricevic, 2006b; as regards phantom limbs in congenital amelia some cases are described by Brugger et al., 2000; Ramachandran & McGeoch,

2007; Saadah & Melzack, 1994; phantom penises are mentioned in Ramachandran & McGeoch, 2008; the case of impossible hues was described by Crane & Piantanida, 1983; Nida-Rumelin & Suarez, 2009; super-saturated colors are discussed in Hurvich & Jameson, 1957; color-synesthesia in the congenitally blind is mentioned but not documented in Ramachandran & Hubbard, 2001.

68 Cooray, Cooray, & Dwyer, 2011.

69 Kanai, Chaieb, Antal, Walsh, & Paulus, 2008; Kar & Krekelberg, 2012; Paulus, 2010.

70 Dobelle, Mladejovsky, & Girvin, 1974; Dobelle & Mladejovsky, 1974; Margalit et al., 2002; Shepherd, Shivdasani, Nayagam, Williams, & Blamey, 2013.

71 The swamp man and a completely congenitally blind subject having a visual phosphene are conceptual analogies. Both of them are supposed to have an experience without any previous contact with light.

72 Gothe et al., 2002; Sayin, 2014.

73 Penfield & Boldrey, 1937; Penfield & Perot, 1963; Penfield & Rasmussen, 1950; Penfield, 1950, 1958, 1972, 1996.

74 Brelén, Duret, Gérard, Delbeke, & Veraart, 2005; Brindley & Lewin, 1968; Chapanis, Uematsu, Konigsmark, & Walker, 1973; Cowey & Walsh, 2000; Dobelle & Mladejovsky, 1974; Gothe et al., 2002; Kammer, Puls, Strasburger, Hill, & Wichmann, 2005; Kanai et al., 2008; Kanai, Paulus, & Walsh, 2010; Kupers et al., 2006; Salminen-Vaparanta et al., 2013; Sayin, 2014; Schutter & Hortensius, 2010.

75 Kennedy & Juricevic, 2006b.

76 Kennedy & Juricevic, 2003, 2006a; Kennedy, 1993; Kennedy & Juricevic, 2006a.

77 Hopkins, 2000; Kennedy, 1993; Lopes, 2002.

78 Stokes, 2005.

79 Amedi, Merabet, Bermpohl, & Pascual-Leone, 2005; L. G. Cohen et al., 1997; Merabet & Pascual-Leone, 2010; Merabet et al., 2009; Pascual-Leone, Amedi, Fregni, & Merabet, 2005; Sadato et al., 1996.

80 Aurora, Welch, & Al-Sayed, 2003; Billock & Tsou, 2012; Kaufman & Solomon, 1992; Pietrobon & Striessnig, 2003; Sacks, 1970; Shams & Plant, 2011.

81 Byrne & Hilbert, 2003; Shapley, 2002; Zeki, 1973.

82 Hering, 1878; Hurvich & Jameson, 1957; Shapley & Hawken, 2011.

83 I happily leave aside the extremely complex issue of what exactly kind of physical properties are colors (Byrne & Hilbert, 2003). Colors are physical properties, no matter how difficult it may be to single them out precisely.

84 Hsieh & Tse, 2006, p. 2251.

85 D. J. Simons & Chabris, 1999; D. J. Simons, 2000; Bonneh et al., 2001; O'Regan, Rensink, & Clark, 1999; Schwitzgebel, 2007.

86 Anstis & Harris, 1975; Bednar & Miikkulainen, 2000; Blakemore & Sutton, 1969; Coltheart, 1971; Craik, 1940; Daw, 1962; Geisler, 1978; Gibson & Radner, 1946; Gilroy & Blake, 2005; Jones, 1972; Kirshfeld, 1999; Köhler & Emery, 1947; Mather, Verstaten, & Anstis, 1998; Mollon, 1974; Phillips, 2013; Sekuler & Ganz, 1963; Shimojo, Kamitani, & Nishida, 2001; Tsuchiya & Koch, 2005; van Boxtel, Tsuchiya, & Koch, 2010; Virsu & Laurinen, 1977a, 1977b; Winawer, Juk, & Boroditsky, 2008; Zaidi, Ennis, Cao, & Lee, 2012.

87 Byrne & Logue, 2009; Fish, 2013; Haddock & Macpherson, 2008; Johnston, 2007; Kalderon, 2011; Martin, 2004; Schellenberg, 2010, 2011.

88 Gerrits, DeHaan, & Vendrik, 1966; Gerrits & Vendrik, 1970; Spillmann, Otte, Hamburger, & Magnussen, 2006.

89 Brugger et al., 2000; Melzack, Israel, Lacroix, & Schultz, 1997; Saadah & Melzack, 1994.

90 McTaggart, 1908; Mellor, 1998; Tooley, 1997.

91 Barbour, 1999; Davies, 2005; Smolin, 2013.

92 Ford, 1974; Power, 2011.

93 Arthurs & Boniface, 2002; Attwell & Iadecola, 2002; Bennett, Wolford, & Miller, 2009; Logothetis & Wandell, 2004; Logothetis, 2008; Schaal, 2005; Uttal, 2001.

94 The attribute *current* has no circularity. It refers simply to the now that contains my writing of the page you are reading in another now.

95 Lennmarken, Bildfors, Enlund, Samuelsson, & Sandin, 2002; Sacks, 2012; Shum, 1999.

96 Jackson & Pargetter, 1977; Salmon, 1969.

97 Ayer, 1956; Chisholm, 1957; Dolev, 2007; Ebersole, 1965; Gale, 1971; Russell, 1927.

98 © David Bowie, *Space Oddity*, 1969.

99 Dowe, 1992, 2007; Reichenbach, 1971.

100 Some authors maintain that a physicalist view is parasitical on a dualistic account of the mind-body problem, as though the only way to outline physicalism were starting from dualism and then denying the mental domain (Montero, 2001, 2013). I do not agree. Rather I agree with Strawson. Strawson, 2008.

101 Edwin Bissell Holt et al., 1910, 1912; Tonneau, 2004.

102 Edwin Bissell Holt, 1914, p. 154.

103 Tonneau, 2004.

104 Valberg, 1992, p. 4.

105 Elsewhere I have used a slightly different terminology with largely the same meaning, derived vs. autonomous representations. Manzotti & Tagliasco, 2001.

106 Dretske, 1977, 1995, 2000; Millikan, 1993.

107 Gallese, 2000; Rizzolatti, Fogassi, & Gallese, 2001.

108 Stoerig, 1996; Weiskrantz, 1990.

109 Koch & Crick, 2001; Reber, 1992.

110 Armstrong, 1961.

111 Doya, 1999; Grieve, Acuña, & Cudeiro, 2000; Ledoux, 2012; Seymour & Dolan, 2008.

112 Driver, Haggard, & Shallice, 2007; Haggard & Libet, 2001; Libet, 1993, 2004.

113 Harnad & Scherzer, 2008.

114 Feyerabend, 1975, p. 46.

INDEX